Deep Learning in Wireless Communications

Haijun Zhang · Ning Yang

Deep Learning in Wireless Communications

Springer

Haijun Zhang
University of Science and Technology
Beijing
Beijing, China

Ning Yang
Institute of Automation
Chinese Academy of Sciences
Beijing, China

ISBN 978-981-97-6316-0 ISBN 978-981-97-6314-6 (eBook)
https://doi.org/10.1007/978-981-97-6314-6

This Springer imprint is published by the registered company Springer Nature Singapore Pte Ltd.
The registered company address is: 152 Beach Road, #21-01/04 Gateway East, Singapore 189721, Singapore

Preface

Intelligent wireless communication plays an important role in the field of communication and other related fields. With the continuous progress and innovation of technology, intelligent wireless communication has become an indispensable part of modern society. It has not only changed the way people live, but also promoted the development of various industries. As an important research direction in intelligent wireless communication, deep learning provides new ideas and methods for system performance optimization, resource management, and user experience. By utilizing deep learning algorithms, intelligent wireless communication systems can automatically learn and adapt to different environments and user needs, thereby providing more personalized and efficient services.

The leap from 5G to 6G means that smart wireless communication has gone through another decade of development. In this new phase, we can expect great changes and innovations in smart wireless communications. For example, higher data rates, lower latency, more connections, and wider coverage will become standard in the 6G era. At the same time, intelligent wireless communication will also be deeply integrated with artificial intelligence, IoT, big data, and other technologies to further promote the digitization and intelligent process of society.

In order to systematically illustrate the development of this field in the past decade and possible future directions, we decided to write a book. This book will cover all aspects of intelligent wireless communication, including technical principles, system design, application scenarios, and future development trends. We will detail the key technologies and standards of 6G, and explore their application prospects in various fields. In addition, we will present application cases of deep learning in intelligent wireless communication and look at new algorithms and technologies that may emerge in the future.

Through this book, we hope to provide readers with a comprehensive and in-depth understanding of intelligent wireless communication opportunities. Whether it is a researcher in a related field, an engineer, or a general reader interested in intelligent wireless communication, you can gain valuable knowledge and inspiration. We believe that this book will become an important reference book in the field of intelligent wireless communication and provide useful guidance for future research and development.

As a scientific research team in the field of intelligent wireless communication, we are committed to the call of The Times, breaking through technical problems and solving existing problems. This book was made possible thanks to the support and contributions of the co-workers.

Beijing, China Haijun Zhang

Contents

1 Introduction to Intelligence Wireless Communication 1
- 1.1 Background on the Evolution of Intelligence Wireless Communication 1
- 1.2 Why Intelligence Wireless Communication Is Important 3
- 1.3 Review the Deep Learning Wireless Communication 7
- 1.4 Summary of Book Content and Chapters 10

2 Cognitive Spectrum Intelligence 13
- 2.1 Deep Spectrum Learning 13
- 2.2 Collaborative Spectrum Sensing 16
- 2.3 Dynamic and Distributed Spectrum Access 19
- 2.4 Supervised Learning-Based Spectrum Sharing 27

3 Learning Resource Allocation Optimization 33
- 3.1 Resource Allocation with Unsupervised Learning 33
- 3.2 DRL for Resource Allocation 35
  - 3.2.1 DRL-Based User Association 39
  - 3.2.2 DRL-Based Channel Assignment and Power Allocation 43
  - 3.2.3 DRL-Based Energy Transfer and Harvesting 49
  - 3.2.4 MADRL-Based Resource Management 53
  - 3.2.5 DRL-Based Task Offloading 58
  - 3.2.6 DRL-Based Load Balancing 62

4 Transmission Intelligence 67
- 4.1 Learning Channel Encoding 67
  - 4.1.1 The Conventional Identification Methods and Its Limitation 67
  - 4.1.2 Deep-Learning-Based Blind Recognition of Channel Code 68

4.2 Channel Estimation and Equalization Methods for Large-Scale MIMO-OFDM Systems 70
4.2.1 Channel Estimation Methods 70
4.2.2 Channel Equalization Methods 72
4.3 Deep Clustering Models 75
4.3.1 Clustering Models Based on Feedforward NNs 75
4.3.2 Deep Clustering Models Based on Autoencoders 76
4.3.3 Deep Clustering Models Based on Generative Models 78
4.3.4 Deep Clustering Model Based on Graph Neural Networks (GNNs) 78
4.4 Adaptive Content Caching 79
4.4.1 The Motivation of DRL for Adaptive Content Caching 79
4.4.2 The Model and Problem Statement of Adaptive Content Caching with DRL 81
4.4.3 DRL for Adaptive Content Caching 81
4.5 DRL for Computing Offloading in Mobile Edge Computing 82
4.5.1 The Motivation of DRL for Computing Offloading in Mobile Edge Computing 82
4.5.2 The Model of Computation Offloading 83
4.5.3 The Algorithmic Procedure of DRL 85
4.6 Beamforming with DRL in MIMO 86
4.6.1 The Motivation of DRL for Beamforming in MIMO 86
4.6.2 The Problem Statement of Beamforming with DRL in MIMO 87
4.6.3 POMDP-Based DRL Algorithm Process for Beamforming 88

5 Learning Traffic and Mobility Prediction 89
5.1 Graph Neural Networks-Based Network Architecture 89
5.1.1 Network Structure 90
5.1.2 GNNs for Communication Networks 94
5.2 Slice Reconfiguration Based on Demand Prediction 96
5.3 Learning Cellular Traffic Prediction 100
5.4 Traffic and Mobility on the Internet of Vehicle 103

6 Software Defined Networking 109
6.1 A Balance Between Flexibility and Manageability 109
6.2 The Role of Intelligent Technology Represented by Machine Learning 112
6.3 Opportunities and Challenges Faced by SDN Under the Integration of Machine Learning 116

7 Security in Wireless Communication 121
7.1 Secure Communications with Supervised Learning 121
7.2 Federated Learning-Based Secure Communications 125

8 6G Driving Applications with Deep Learning 133
8.1 Application Scenarios and Challenges 133
8.2 Enabling Deep Learning Technologies 137
8.3 New Paradigm Shifts 140

Acronyms

1G	The First-Generation Cellular Network
3GPP	The Third-Generation Partnership Programme
AC	Actor-Critic
ACC	Accuracy
ACSS	Attack-aware Collaborative Spectrum Sensing
AE	Auto-encoder
AGAE	Adversarial Graph Auto-encoders
AI	Artificial Intelligence
AP	Access Points
AR	Augmented Reality
Bi-RNN	Bidirectional RNN
BS	Base Stations
CatGAN	Categorical Generative Adversarial Network
CDMA	Code Division Multiple Access
CNNs	Convolutional Neural Networks
CR	Cognitive Radio
CRSN	Collaborative Spectrum Sensing Model
CSS	Collaborative Spectrum Sensing
DAC	Deep Adversarial Clustering
DEC	Deep Embedded Clustering
DEN	Deep Embedding Network
DNN	Deep Neural Network
DQN	Deep Q-network
DRL	Deep Reinforcement Learning
DSA	Dynamic Spectrum Access
DSC-Nets	Deep Subspace Clustering Networks
eMBB	Enhanced Mobile Broadband
FC	Fusion Center
GAE	Graph Auto-encoder
GAEs	Graph Auto-encoders
GAN	Generative Adversarial Networks

GAT	Graph Attention Network
GATs	Graph Attention Networks
GCNs	Graph Convolutional Networks
GGNs	Graph Generative Networks
GMVAE	Gaussian Mixture VAE
GNNs	Combining Graph Neural Networks
GPRS	General Packet Radio Service
GRN	Graph Recurrent Network
GRNs	Graph Recurrent Networks
GRU	Gated Recurrent Unit
HST	High-Speed Train
IDS	Intrusion Detection Systems
InfoGAN	Information Maximizing Generative Adversarial Network
IoT	Internet of Things
LFU	Least Frequently Used
LMMSE	Linear Minimum Mean Square Error
LRU	Least Recently Used
LS	Least Squares
LSTM	Long Short-Term Memory
LTE	Long-Term Evolution
MADRL	Multi-agent Deep Reinforcement Learning
MARL	Multi-agent Reinforcement Learning
MCO	Mobile Computing Offloading
MDP	Markov Decision Process
MEC	Mobile-Edge Computing
MIMO	Multiple-Input Multiple-Output
ML	Maximum Likelihood
MMSE	Minimum Mean Square Error
mMTC	Massive Machine Type Communications
mmwaves	Millimeter Waves
NFV	Network Function Virtualization
NMI	Normalized Mutual Information
NN	Nearest Neighbor
NNs	Neural Networks
No	Noise Spectral Density
OFDM	Orthogonal Frequency Division Multiplexing
OSPF	Open Shortest Path First
PF	Proportional Fair
PG	Policy Gradients
POMDP	Partially Observable Markov Decision Process
PPO	Proximal Policy Optimization
PRB	Physical Resource Block
PU	Primary User
QoS	Quality of Service
RANs	Radio Access Networks

RF	Radio Frequency
RFADC	Recurrent Framework Agglomerative Deep Clustering
RL	Reinforcement Learning
RNNs	Recurrent Neural Networks
RSS	Received Signal Strength
SARSA	State-Action-Reward-State-Action
SDN	Software-Defined Networking
SDR	Software-Defined Radio
SD-RANs	Software-Defined RANs
SINR	Signal-to-Interference-Plus-Noise Ratio
SLAs	Service-Level Agreements
SMS	Short Message Service
SNR	Signal-to-Noise Ratio
SSL	Secure Socket Layer
STFT	Short-Time Fourier Transform
STL	Seasonal Decomposition of Time Series
SU	Secondary Users
SVMs	Support Vector Machines
TD	Temporal Difference
TRPO	Trust Region Policy Optimization
UEs	User Equipments
URLCC	Ultra-Reliable and Low Latency Communications
V2I	Vehicle to Infrastructure
VaDE	Variational Deep Embedding
VAE	Variational Auto-encoders
VLANs	Virtual Local Area Networks
VR	Virtual Reality
ZF	Zero Forcing

Chapter 1
Introduction to Intelligence Wireless Communication

1.1 Background on the Evolution of Intelligence Wireless Communication

Intelligent wireless communication is the benchmark of modern communication, but its development is not achieved overnight. The development background of intelligent wireless communication mainly has the following four points.

With the rapid expansion of modern population, the rapid progress of science and technology, mobile Internet, Internet of things (IoT), machine learning and other technologies are booming. This makes smart devices popular, application scenarios richer, and therefore spawned the need for high-speed, reliable, effective wireless communication technology. As the global population continues to grow, more and more people and devices need to be connected. The rise of mobile Internet enables people to access the Internet anytime, anywhere through smartphones, tablets and other mobile devices to enjoy rich information and services. The IoT connects various smart devices and sensors with the Internet to realize intelligent interaction and information sharing between devices. The development of technologies such as machine learning and artificial intelligence (AI) enables devices to learn and adapt to the needs of users to provide more intelligent services. With the continuous progress and popularization of these technologies, people's demand for wireless communication technology is getting higher and higher. They want to be able to enjoy high-speed, reliable and efficient wireless connectivity, whether through mobile devices or through various IoT devices. Whether it is watching high-definition videos, online games, telecommuting, or transferring data between IoT devices, wireless communication technology needs to be able to provide stable and fast connections to meet users' requirements for data transmission rates and latency. In addition, with the rapid development of smart cities, smart transportation, smart home and other fields, wireless communication technology is also playing an increasingly important role. People want to achieve a more convenient lifestyle through smart devices, and wireless communication technology needs to provide reliable connections to support

H. Zhang and N. Yang, *Deep Learning in Wireless Communications*,
https://doi.org/10.1007/978-981-97-6314-6_1

the interconnection between smart devices. For example, controlling home devices through smartphones, real-time monitoring of urban traffic conditions, telemedicine, etc., all require wireless communication technology to provide fast, reliable connections and stable data transmission. Therefore, with the rapid expansion of people around the world and the progress of science and technology, the vigorous development of mobile Internet, IoT, machine learning and other related technologies has spawned a huge demand for high-speed, reliable and efficient wireless communication technology. Whether it is to meet users' requirements for data transmission rates and latency, or to support connectivity between smart devices, the development of wireless communication technology will continue to innovate, driven by changing social needs and technological advances.

The bottleneck of the original wireless communication technology, such as the shortage of spectrum resources. With traditional methods, limited spectrum resources cannot meet the growing demand. In response to this problem, intelligent wireless communication has emerged as a solution to enable more efficient and flexible use of spectrum resources by leveraging cognitive spectrum intelligence and other technologies. Spectrum resource is the key resource of wireless communication, which is the electromagnetic frequency range required for transmitting data and signals. However, due to the limited spectrum resources, the traditional wireless communication technology is faced with serious challenges. In the past, people mainly relied on fixed spectrum resources allocated to various communication services to meet communication needs. However, with the continuous growth of wireless communication demand, the traditional spectrum allocation mode becomes more and more unsuitable for the actual demand, resulting in a relatively low utilization rate of spectrum resources and low efficiency. In order to solve this problem, intelligent wireless communication technology comes into being. Intelligent wireless communication is committed to improving the efficiency and flexibility of spectrum resources to meet the growing demand for wireless communication. Among them, cognitive spectrum intelligence technology is a key technology, which enables wireless devices to sense and understand the spectrum environment in which they are located. By utilizing cognitive spectrum intelligence technology, wireless devices can quickly detect, evaluate and utilize available spectrum resources to adapt to real-time communication needs.

The third point is the emergence of programmable wireless devices and machine learning technology in modern times, that is, intelligent technology. Programmable wireless devices refer to wireless devices with programmable functions, which can be flexibly configured and optimized according to different needs and scenarios. Through software-defined methods, programmable wireless devices can dynamically adjust communication frequency, signal selection, power control and other parameters according to the actual situation to meet different wireless communication requirements. This flexible programming feature enables wireless devices to better adapt to changing communication environments and provide higher and more reliable wireless connections. Machine learning technology plays a key role in intelligent wireless communication. By utilizing a learning algorithm, wireless devices can learn from large amounts of data and automatically optimize communication performance. In the past, online networks needed to be manually configured and optimized, but the introduction of machine learning technology allows wireless networks to automatically learn and make decisions for more efficient and stable communication services. Machine learning technology can analyze historical data and real data by intelligent learning algorithms, continuously optimize the performance of line network, improve spectrum utilization, reduce transmission delay and provide better user experience. The combination of programmable wireless devices and machine learning technology provides good conditions for the development of intelligent wireless communication. The scalable wireless device provides the ability of flexible configuration and optimization so that the wireless communication network can be dynamically adjusted according to the actual demand, and improve the communication efficiency. The machine learning technology can make the communication network adapt and improve continuously through autonomous learning and optimization to provide better communication performance.

Finally, there is the push of 5G technology. 5G technology has the characteristics of greater bandwidth, lower latency and more connected devices, so as a new era of intelligent wireless communication technology, 5G technology will promote the development of related fields. At the same time, the large bandwidth characteristics of 5G technology will greatly improve the transmission capacity of wireless communication, the low latency characteristics of 5G technology will achieve faster data transmission response speed, and the more connected device characteristics of 5G technology will promote the wide application of the IoT and smart devices. In addition, it is able to connect more devices at the same time, through higher spectral efficiency and multi-user multiple-input multiple-output technology, to achieve a large number of devices at the same time seamless communication. This makes it more convenient for IoT devices to realize interconnection and promote the rapid development of smart home, smart city, intelligent transportation and other fields. With the continuous evolution and popularization of 5G technology, intelligent wireless communication will usher in broader and promising development opportunities. In summary, the development of intelligent wireless technology in the new era, the bottleneck caused by the shortage of spectrum resources, the research and development of 5G technology and the increasing demand for wireless communication have jointly promoted the development of intelligent wireless communication, and provided more possibilities and opportunities for intelligent wireless communication in the future.

1.2 Why Intelligence Wireless Communication Is Important

The importance of intelligent wireless communication is mainly reflected in its wide application, and it has an important position in many related fields. From the Industrial IoT, autonomous driving, robotics, virtual reality (VR) and augmented reality

(AR), healthcare, to the haptic Internet, the application and importance of intelligent wireless communication is everywhere.

First of all, the IoT can realize the connection and communication between devices through intelligent wireless communication, achieve automated production and management, improve production efficiency, reduce costs, and improve product quality. IoT is a network that connects physical devices, sensors and computers to each other through wireless communication technology, featuring large-scale connectivity and mobility support.

The popularity of smart sensors and IoT gateways has enabled large-scale connectivity between industrial devices. The sensors are able to collect the status information of the device and transmit it in real-time to the IoT gateway, which is responsible for processing and forwarding this data. In this way, industrial equipment can be interconnected and data can be shared and analyzed instantly, thereby improving production efficiency and product quality. Smart technology supports the mobility of industrial equipment. Through the introduction of intelligent robots and automatic navigation technology, industrial equipment is able to move and position itself autonomously. The intelligent robot is equipped with various sensors and navigation systems, which can accurately perceive the surrounding environment and plan the optimal path, improving the flexibility and adaptability of the device. At the same time, intelligent technology can also achieve collaborative operation between devices to improve the efficiency and automation level of industrial production. Intelligent technology also supports the connectivity and data interaction between industrial equipment and cloud platforms. By uploading equipment data to the cloud platform, industrial enterprises can achieve remote monitoring and management of equipment status and production processes. Intelligent analysis algorithms on the cloud platform can conduct in-depth processing and mining of data to help enterprises find problems and optimize production processes.

In the field of autonomous driving, intelligent wireless communication is the key to realising information interaction and collaboration between vehicles. Autonomous vehicles can share real-time road conditions, vehicle status and other information through wireless communication to make real-time collaborative decisions, optimize traffic flow and improve road safety. Intelligent sensors and cognitive systems ensure the high reliability of autonomous driving. Autonomous vehicles are equipped with a variety of sensors, such as lidar, cameras, and ultrasonic sensors, which can obtain environmental information such as road conditions and obstacles in real-time. Through the processing and analysis of sensor data, the cognitive system can accurately perceive and understand the surrounding environment, and make corresponding decisions. With these intelligent technologies, autonomous vehicles can achieve accurate location awareness, obstacle avoidance and traffic accident avoidance, ensuring driving safety and high reliability. And smart technology provides mobility support for autonomous driving. By introducing positioning systems, map data, and navigation algorithms, autonomous vehicles are able to achieve accurate positioning and path planning. The positioning system can accurately locate the location of the vehicle, the map data provides information such as road conditions and road speed limits, and the navigation algorithm can optimize the driving path, taking into account the traffic flow, vehicle speed and other factors to achieve fast and efficient travel. At the same time, intelligent technology can also realize automatic parking, remote scheduling and other functions, providing convenient mobility support.

For robots, the development of intelligent technology provides technical support for robots to achieve high energy efficiency and mobility support, making robots make major breakthroughs in data interaction and communication. As an autonomous system capable of performing tasks independently, robots usually need to interact with the environment, perceive external information, and make decisions based on the information obtained. Intellisense and learning technology can improve the energy efficiency of robots. The robot is equipped with various sensors, such as cameras, infrared sensors, etc., and through real-time perception and recognition of the environment, it can optimize the use of resources according to actual needs. For example, robots can intelligently adjust brightness according to light conditions, adapt speed according to object size, reduce energy consumption and improve energy efficiency. Intelligent path planning and navigation technologies provide support for the efficient movement of robots. Equipped with a navigation system, combined with map data and planning algorithms, the robot can automatically plan the optimal path and avoid obstacles. In addition, intelligent path planning technology can dynamically adjust the path based on real-time traffic information, ensuring that the robot travels efficiently and safely during the movement. In addition, the intelligent charging and energy management system enables the robot to have highly energy efficient charging characteristics. Through the intelligent charging system, the robot can autonomously adjust the charging speed and timing according to the battery capacity and usage, minimizing energy waste.

In the field of VR and AR, intelligent technology has played an important role in promoting high broadband and high-speed transmission. For example, 5G is the fifth-generation mobile communication technology, which has the characteristics of high bandwidth and low latency, and provides high-speed transmission support for VR and AR applications. The deployment and popularization of 5G networks allow users to download and upload large amounts of data faster, resulting in smoother VR and AR experiences. At the same time, intelligent technology can monitor and optimize the network in real-time to provide more stable and high-speed transmission services. Through intelligent network management and optimization mechanisms, network bandwidth can be adjusted according to real-time data traffic, providing the best transmission quality and user experience. In addition, data compression and optimization algorithms are able to compress this data to a smaller size to reduce bandwidth requirements and maintain low latency during transmission, providing a high-quality transmission experience.

In the healthcare sector, smart wireless communication is widely used in telemedicine, health monitoring and smart medical devices. It ensures that only legitimate medical personnel have access to sensitive medical data by using secure encryption protocols, authentication, and access control mechanisms, preventing data from being accessed or tampered with by unauthorized personnel. At the same time, intelligent technology in the medical field can achieve high-bandwidth transmission

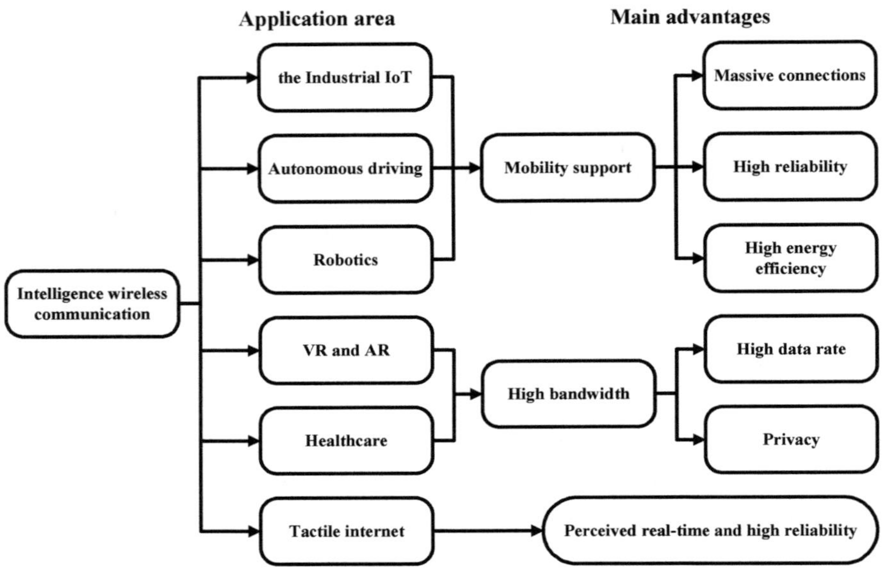

Fig. 1.1 Application scenarios and advantages of intelligent technology in wireless communication

to meet the needs of large-capacity and high-speed transmission in the medical field. Using cloud computing and distributed systems, intelligent technologies can provide high-bandwidth information transmission in large-scale, complex medical data processing and analysis. The elastic and resource-sharing nature of cloud computing can effectively support the demand for large-scale data processing and transmission in healthcare. This is conducive to promoting the real-time transmission of medical data, promoting the development of telemedicine, and improving the quality and efficiency of medical services.

Haptic Internet is an emerging concept that enables the transmission and sharing of haptic and force feedback through intelligent technology. The perceptual data in the tactile Internet usually contains a lot of information, such as tactile images, videos, etc., which needs to be transmitted through high-bandwidth communication. In the haptic Internet, high-speed sensors and devices are needed to sense, collect and transmit haptic information. Intelligent technology can continuously improve and develop high-rate sensor technology to improve the efficiency of data acquisition and transmission.

In general, intelligent wireless communication has important applications and influence in the fields of industry, transportation, healthcare and interactive entertainment. It not only connects people and devices, devices and devices, but also enables more efficient information transmission and collaborative decision-making, thus promoting the progress and development of society (Fig. 1.1).

1.3 Review the Deep Learning Wireless Communication

Wireless communication plays a vital role in our modern society, enabling seamless connectivity and communication between individuals, devices, and systems.

The utilization of deep learning in wireless communication has emerged as a prominent research hotspot and trend due to several key factors. Deep learning, as a subfield of AI, has showcased remarkable capabilities in various domains, including computer vision, natural language processing, and speech recognition. Its potential to revolutionize wireless communication has sparked intense research interest for the following reasons: wireless communication systems are becoming increasingly complex, with a growing number of devices, diverse communication protocols, and rapidly evolving network architectures. Deep learning techniques, such as neural networks (NNs), can effectively handle this complexity by automatically learning hierarchical representations from vast amounts of data. This enables the extraction of hidden patterns, optimization of system parameters, and improved performance in complex wireless environments. Deep learning offers the potential for intelligent resource management in wireless networks. By leveraging large-scale data analysis and advanced machine learning techniques, deep learning algorithms can optimize resource allocation, power control, and network scheduling. This facilitates efficient utilization of network resources, reduced latency, improved energy efficiency, and overall network performance. Cognitive radio (CR) networks, which aim to dynamically access and utilize available spectrum, can greatly benefit from deep learning. Deep learning models can analyze radio environment parameters, detect spectrum holes, and enable efficient spectrum sharing and allocation. This allows for improved spectrum utilization, increased network capacity, and enhanced coexistence between different wireless systems. Deep learning is expected to play a significant role in next-generation wireless technologies, such as 5G, beyond 5G, and IoT networks. These technologies require intelligent, adaptive, and self-optimizing systems that can handle massive data flows and diverse communication scenarios. Deep learning can provide the necessary tools and techniques to effectively address the complexities and challenges posed by these emerging wireless technologies. This book aims to explore the relationship between deep learning and wireless communication, focusing on the key technologies involved and addressing the challenges and future directions of integrating deep learning into wireless communication systems.

Deep learning plays a crucial role in wireless communication systems by enabling intelligent data processing, decision-making, and resource allocation. Its applications include:

- Signal processing. Channel processing means that deep learning algorithms can learn to extract relevant features from the received signal, acting on noise, interference, and channel fading effects. This helps improve signal quality and enhances detection, demodulation, and decoding performance.
- Channel estimation and prediction. Deep learning models can accurately estimate and predict wireless channel characteristics based on historical data. This helps

to design adaptive communication strategies and optimize system performance under dynamic channel conditions.

- Interference management. Deep learning techniques effectively simulate and mitigate interference in wireless communication systems. By intelligently receiving signals and recognizing interference patterns, deep learning algorithms can improve the reliability and throughput of communications.
- CR networks. Deep learning can realize intelligent spectrum sensing, dynamic spectrum access (DSA) and efficient spectrum allocation in CR networks. By learning from historical data and environmental information, deep learning models can make informed decisions to achieve optimal spectrum utilization and avoid interference.

In summary, deep learning plays a crucial role in wireless communication systems, offering applications such as signal processing, channel estimation, interference management, and CR networks. Compared to traditional methods, deep learning provides advantages through end-to-end processing, adaptability, handling complexity, and robustness. By leveraging deep learning techniques, wireless communication systems can benefit from improved performance, enhanced reliability, and optimized resource allocation.

Deep learning techniques have been widely adopted in wireless communication systems, leveraging various key technologies to enhance performance and enable intelligent decision-making. The following are commonly used key technologies in deep learning for wireless communication, including NNs, CNN, and RNN:

1. Neural Networks: NNs are the foundation of deep learning and play a vital role in wireless communication applications. They consist of multiple interconnected layers of artificial neurons that can learn and extract patterns from input data. NNs can be trained using backpropagation algorithms to optimize their weights and biases, enabling accurate data processing and decision-making.

2. Convolutional Neural Networks (CNNs): CNN are primarily used for image and signal processing tasks in wireless communication systems. They excel at capturing spatial dependencies and extracting relevant features from input data. Their key components include convolutional layers, pooling layers, and fully connected layers. CNNs are trained to automatically learn hierarchical representations, enabling efficient feature extraction and classification of signals in wireless communication.

3. Recurrent Neural Networks (RNNs): RNNs are designed to process sequential data, making them well-suited for wireless communication tasks involving time-varying or sequential signals. RNNs utilize a feedback mechanism, allowing them to capture temporal dependencies in input data. This makes them effective in tasks such as channel estimation, speech recognition, and DSA. Popular variants of RNNs include Long Short-Term Memory (LSTM) and Gated Recurrent Unit (GRU) networks.

4. Reinforcement Learning (RL): RL is a subfield of deep learning that focuses on learning optimal actions through interaction with an environment. In wireless communication systems, RL can be utilized for CR networks, adaptive resource

allocation, and dynamic spectrum management. By learning from rewards and penalties, RL algorithms can optimize system performance and make intelligent decisions based on changing network conditions.

5. Transfer Learning: Transfer learning allows knowledge learned from one domain or task to be transferred and applied to another related domain or task with limited labeled data. In wireless communication, where labeled data can be scarce, transfer learning techniques can be used to leverage pre-trained models or knowledge learned from related tasks, improving the efficiency and accuracy (ACC) of deep learning algorithms.

6. Federated Learning: Federated learning realizes localized data processing and decentralized model training by delegating the model training to the terminal equipment. Through multi-level neural network models and large-scale training data, deep learning can extract complex features from data to achieve efficient wireless communication system optimization.

These key technologies play crucial roles in enabling deep learning capabilities in wireless communication systems. By leveraging NNs, CNNs, RNNs, RL, and transfer learning, deep learning algorithms can efficiently process wireless signals, extract meaningful features, optimize resource allocation, and improve overall system performance.

Deep learning in wireless communication faces several challenges, including computational complexity, insufficient training data, and the need for addressing real-time constraints. However, these challenges also present opportunities for future research and development. The following analysis highlights the challenges and provides insights into the potential future directions of deep learning in wireless communication challenges:

Computational Complexity: Deep learning models often involve complex computations and require significant computational resources. High computational complexity can limit the real-time processing capability of deep learning algorithms in resource-constrained wireless communication devices. Developing efficient algorithms and optimizing hardware to reduce computational complexity is an ongoing challenge.

Insufficient Training Data: Training deep learning models typically requires large amounts of labeled data, which may be scarce or expensive to acquire in wireless communication scenarios. Obtaining sufficient and diverse training data for specific wireless communication tasks is a challenge. Techniques like transfer learning and data can help overcome this challenge by leveraging pre-trained models and generating synthetic training data.

Real-time Constraints: Wireless communication systems operate in real-time environments with stringent delay requirements. The latency introduced by the training and inference processes of deep learning algorithms may exceed these real-time constraints. Developing real-time deep learning algorithms and designing efficient hardware architectures to accelerate model training and inference is an ongoing challenge. Future Directions: Edge computing and federated learning. Moving computation and decision-making closer to the edge of the network can alleviate the computational complexity and latency challenges in deep learning for wireless communication. Edge computing and federated learning enable distributed training and inference on edge devices, reducing reliance on centralized servers and improving privacy and energy efficiency.

Transfer Learning and Semi-Supervised Learning: To address the challenge of limited labeled training data, transfer learning and semi-supervised learning techniques can be further explored in wireless communication. Pre-training models on large-scale data from related tasks or domains and utilizing unsupervised learning methods to leverage unlabeled data can enhance the performance and generalization capability of deep learning models.

Explainable and Interpretable Deep Learning: The lack of interpretability in deep learning models can pose challenges in wireless communication systems, where transparency and interpretability are essential. Research on developing explainable deep learning models, interpretability metrics, and visualization techniques can enable better understanding and trust in the decision-making processes of deep learning algorithms in wireless communication.

RL and Autonomous Networks: Further exploration of RL techniques can enable the development of autonomous wireless communication networks. RL algorithms can dynamically adapt to changing network conditions, optimize resource allocation, and improve network efficiency. Autonomous networks empowered by deep reinforcement learning (DRL) can lead to self-organizing, self-optimizing, and self-healing wireless communication systems.

Hybrid Approaches: Hybrid approaches that combine deep learning with traditional signal processing methods can leverage the strengths of both approaches for enhanced performance in wireless communication systems. Combining deep learning techniques with domain knowledge, physical models, and optimization algorithms can provide a comprehensive and effective solution for addressing the challenges in wireless communication.

In conclusion, while deep learning in wireless communication faces challenges such as computational complexity and limited training data, there are promising future directions to explore. Edge computing, transfer learning, explainable deep learning, RL, and hybrid approaches are among the potential research areas that can further advance the application of deep learning in wireless communication, enabling more intelligent, efficient, and reliable wireless communication systems in the future.

1.4 Summary of Book Content and Chapters

The main contents of this book are as follows: PART II mainly deals with the content related to cognitive spectrum intelligence, such as deep spectrum sensing, Collaborative Spectrum Sensing (CSS), DSA, etc. PART III mainly discusses the optimization of learning resource allocation, including the use of supervised learning and unsupervised learning for power allocation and resource allocation, and the use of RL for user association, channel allocation and energy transfer resource management.

PART IV discusses topics related to transmission intelligence, such as learning channel coding, decoding and forwarding, deep clustering networks, etc. It also covers computational offloading using RL in mobile edge computing and beamforming using DRL. PART V discusses learning traffic and mobility forecasting, including network architecture based on graph NNs, slicing reconfiguration based on demand forecasting, learning cellular traffic forecasting, and traffic and mobility for drones and IoT vehicles. PART VI deals with security issues in wireless communication, including secure communication using methods such as supervised learning and federated learning. PART VII focuses on 6G driven applications based on deep learning, including application scenarios, industry verticals, and new paradigm shifts such as air-ground-sea integrated communication networks and sub-6 GHz, millimeter wave, and terahertz technologies. PART VIII summarizes the content of the book and gives the research agenda and the direction of future work. Through the arrangement of these subheadings, the logical structure of the whole chapter is relatively clear, and the importance, technical application and research direction of intelligent wireless communication are shown in turn.

All in all, the book covers several key topics in intelligent wireless communications, from cognitive spectrum intelligence to resource allocation, transmission intelligence, traffic and mobility prediction, and security and 6G applications. This comprehensive perspective allows readers to systematically understand all aspects of intelligent wireless communication and their relationships. Each chapter provides specific content and domain segmentation, from perception, access, and distribution to transmission, prediction, and security. This depth and breadth allow the reader to delve into the details of each topic and gain a comprehensive understanding. At the same time, the book covers many practical application scenarios and solutions, such as mobile edge computing, IoT vehicles, and drone communications. These cases and examples can provide practical reference for readers and help them understand the application and significance of intelligent wireless communication in practice. Finally, at the end of the book, the authors provide a future research agenda and open questions. These outlooks provide a guide to explore deeper questions and develop new research, encouraging readers to make innovations and contributions in the field of intelligent wireless communications. In summary, this book not only has comprehensive content coverage, but also provides information on cutting-edge technologies, practical applications and future development directions, which can help readers deeply understand the field of intelligent wireless communication, and provide valuable guidance and reference for further research and practice.

Chapter 2
Cognitive Spectrum Intelligence

2.1 Deep Spectrum Learning

As 6G technology research gradually matures and the IoT rapidly gains momentum, the demand for wireless spectrum has significantly increased. Spectrum sensing plays a crucial role in this context. In this section, we will focus on introducing the technology of deep spectrum sensing. Deep spectrum sensing refers to the application of deep learning techniques, and it is a critical function in CR and DSA systems, where devices need to detect and utilize available radio frequency (RF) spectrum bands opportunistically and efficiently. Traditional spectrum sensing methods often rely on signal processing techniques and statistical analysis to detect the presence of primary users or other wireless devices in a specific frequency band. Deep spectrum sensing leverages DNNs to improve the ACC and robustness of this detection process. The implementation of deep spectrum sensing requires the assistance of relevant algorithms, and different algorithms have different characteristics. Below, we introduce some different deep learning methods:

In CR, most spectrum sensing algorithms are model-based, the model-based spectrum sensing algorithm is a method that utilizes mathematical models and signal processing techniques to understand and perceive the wireless radio spectrum. This algorithm typically employs machine learning or statistical methods to infer the characteristics and properties of different signals in the radio spectrum by building models. However, its detection performance heavily relies on the ACC of the assumed statistical models. If the established models are not accurate enough or cannot adapt well to the changing real-world wireless environment, the algorithm's performance may be impacted. For instance, in the case of model-based spectrum sensing algorithms, assumptions about the model may involve some statistical descriptions of signal characteristics, noise distribution, or spectrum occupancy patterns. If the data in the actual environment does not align with these assumptions, the algorithm's performance may be affected. Based on this observation, researchers have proposed additional algorithms.

H. Zhang and N. Yang, *Deep Learning in Wireless Communications*,
https://doi.org/10.1007/978-981-97-6314-6_2

In the realm of deep spectrum sensing, we can employ deep learning algorithms based on CNNs. In comparison to model-based spectrum sensing algorithms, their proposed deep learning methods are data-driven, accomplishing sensing without the need for probability models or knowledge of Primary User (PU) activity patterns. They can be trained on a substantial amount of annotated data to learn features and patterns, enabling effective classification or regression on unlabeled data thereafter. This data-driven approach showcases the excellence of CNN in fields such as image processing and language, as they automatically learn intricate feature representations from data without the necessity of manually designing feature extractors. In the context of spectrum sensing, CNN exhibit advantages in low Signal-to-Noise Ratio (SNR) environments. They autonomously learn features that adapt to low SNR conditions, capturing subtle differences between signals and noise effectively. Through convolutional layers, CNN efficiently conduct local perception, aiding in extracting local structures and patterns within signals to counteract the impact of noise. While CNN are data-driven, for more complex spectrum sensing tasks, such as those in wireless communication, specialized algorithms and models like RNN are often employed. RNNs are designed to better capture the temporal variations and spectral characteristics of signals.

The term methods based on RNN refers to a category of neural network approaches that utilize the RNN structure for modeling and processing. RNN is a neural network architecture specifically designed for handling sequential data, possessing the ability to capture temporal dependencies within sequences. In methods based on RNNs, the core idea of RNN involves introducing a recursive structure, allowing the network to consider previous information when processing each time step. Its fundamental components include recursive connections, hidden states, weight parameters, and time steps. RNNs can be applied to predict future values in time series data, such as received signals. This is applicable in various domains, including speech recognition, stock price prediction, weather forecasting, etc. By learning patterns and dependencies in time series data, RNNs can predict future values given past input sequences. Subsequently, covariance matrices can be used for spectrum sensing. A covariance matrix describes the relationship between two or more random variables. In signal processing and communication, covariance matrices are commonly used to analyze the statistical properties of signals. In spectrum sensing, covariance matrices can be applied for signal separation, spectrum estimation, spatial spectrum sensing, etc. Particularly in multi-antenna systems, analyzing the covariance matrix can optimize the performance of communication systems. In multi-antenna communication systems, analyzing the covariance matrix of received signals allows for signal separation, improving the ACC of modeling multipath channels. This helps the system better understand the effects of multipath propagation and optimize the decoding process of received signals. In multi-antenna systems, by analyzing the covariance matrix of received signals, signal separation can be performed, enhancing the reliability and performance of communication systems in situation involving multipath and channel fading. This contributes to increased spatial diversity and improved system reliability and performance in situations involving multipath and channel fading.

In the field of spectrum sensing, autoencoders can be employed to learn effective spectrum representations. By leveraging the structure of autoencoders to extract behavioral features from raw signals, training enables the acquisition of compact data representations for better capturing spectral characteristics. In this approach, recursive autoencoders or RNNs can be utilized to model dynamic spectrum changes, thereby adapting to the time-varying nature of the channel. The process of extracting raw signals typically involves signal processing and feature engineering, encompassing time-domain features, frequency-domain features, time-frequency-domain features, advanced features, and nonlinear features. Short-time Fourier transform (STFT) is a technique applicable to frequency-domain and time-frequency-domain features.

STFT is a method for time-frequency analysis of signals, providing high resolution in both time and frequency domains. It applies a time window function to the signal and calculates the magnitude spectrum within each window. Specifically, STFT calculates the signal's spectrum by shifting the window along the time axis and applying a Fourier transform to each segment of the signal within these time windows. These short-time signal segments significantly enhance the ACC of frequency domain analysis, enabling the extraction of time-frequency features of the signal over short durations, allowing for a better analysis of signal characteristics like frequency, phase, and magnitude. In the frequency domain, each window's Fourier transform produces a spectrum. In the time domain, by sliding the window function, spectrum information for the signal at different time intervals can be obtained.

$$X(t, \omega) = \int_{-\infty}^{\infty} x(\tau) \cdot w(\tau - t) \cdot e^{-j\omega\tau}\, d\tau \tag{2.1}$$

$X(t, \omega)$ represents the STFT result at time t and frequency ω. $x(\tau)$ denotes the input signal, a function of time τ in this formula. $w(\tau - t)$ represents the window function sliding along the time axis, selecting a local portion of the signal for the Fourier transform. ω is the angular frequency, representing a unit of frequency. $e^{-j\omega\tau}$ is the complex exponential function used in the Fourier transform.

Therefore, STFT provides a precise analysis tool in both time and frequency domains. It finds extensive applications in fields such as signal processing and speech recognition. In the context of spectrum sensing, STFT can be used to extract features of signals in both time and frequency, improving signal identification. Compared to other traditional spectrum sensing methods, using STFT for spectrum analysis offers better performance and higher ACC.

In general, the application of DRL in the spectrum sensing domain can enhance the efficiency, capacity, and reliability of wireless communication systems, enabling them to better adapt to complex and dynamic spectrum environments.

2.2 Collaborative Spectrum Sensing

CSS in wireless communication aims to efficiently manage spectrum resources by enabling different devices to work together, enhancing efficiency, and reducing interference. Through methods like spectrum sensing, sharing, and dynamic allocation, collaborative sensing increases opportunities for spectrum acquisition. Nodes distributed in various locations during collaborative sensing provide spatial diversity, while collecting data at different times enhances the capture of dynamic spectrum changes. Sharing resources and information among nodes optimizes resource utilization and improves spectrum acquisition efficiency. In the contexts of 5G, IoT, and wireless communication, CSS is crucial for addressing the challenges of limited spectrum resources, optimizing network performance, and meeting diverse communication demands.

The utilization of machine learning methods can enhance CSS. Employing CNNs for spatial feature extraction enables the capture of local patterns in the spectrum, such as signal edges and textures. RNN are effective in handling the time-series nature of spectrum data, capturing changes in signals over time. Transfer learning facilitates knowledge sharing across diverse environments. Online learning algorithms dynamically adapt to changing spectrum conditions. Robustness is improved through the integration of information from multiple sources, and self-supervised learning enhances generalization. Consideration of RL optimizes decision-making strategies within the system. Tailoring these methods to specific task requirements and application scenarios can provide more accurate and efficient solutions for CSS.

Spatial and temporal collaboration is also one of the research directions in CSS. Spatial collaboration focuses on effectively organizing and coordinating resources, nodes, or activities in space to achieve better efficiency and coverage. In fields such as sensing networks, IoT, and communication systems, spatial collaboration may involve the deployment of sensing nodes, the layout of device locations, and the planning of wireless communication networks. Through rational spatial collaboration, resources can be maximized, ensuring comprehensive sensing or communication coverage within the designated area. Temporal collaboration, on the other hand, concentrates on coordinating and scheduling resources, information, or activities at different time points to adapt to dynamic environmental changes. In scenarios like spectrum management and dynamic resource allocation, temporal collaboration can optimize resource utilization at different time points. This includes collecting data at specific moments, adjusting spectrum allocation, dynamically tuning system parameters, and meeting the demands and environmental conditions of different time periods. Deep learning can optimize the deployment of sensing nodes. Using deep learning models, environmental features of the sensing area, including terrain, building distribution, and obstacle positions, can be learned and extracted. Deep learning models can learn advanced environmental representations from sensor data, aiding in a better understanding of the structure of the sensing area. In terms of temporal collaboration, deep learning can be applied to analyze the dynamic changes in spectrum data. Deep learning algorithms, such as CNN and RNN, can be used to analyze spectrum data. These algorithms can learn complex spectrum patterns, trends, and changes, enabling the system to better comprehend the state of the spectrum at different time points. The system can adjust spectrum resource allocation, channel selection, and other parameters based on the output of deep learning models to optimize communication performance.

The Collaborative Spectrum Sensing Model (CRSN) is a form of implementation or technological framework of CSS. The CRSN model aims to achieve efficient CSS and rapid detection of malicious network nodes while ensuring security and energy efficiency. Network CSS technology based on topological clustering and channel state information model is typically a method that combines network topology and channel state information to effectively utilize multiple nodes in the network for spectrum environment sensing. By employing topological clustering, the system can form a distributed sensing structure, assigning sensing tasks to different nodes or clusters, making the system more flexible and adaptable to various network topologies and environments. Through topological clustering and the channel state information model, the system can establish a trust model for nodes. This helps in identifying and verifying legitimate nodes, assessing their reliability in sensing tasks. Through trust modeling, the system can increase confidence levels in node identities, reducing the risk of malicious attacks. Furthermore, introducing elements of deep learning can further enhance the system's performance, particularly in the following aspects: By integrating deep learning models, the system can intelligently leverage multiple nodes in the network for the extraction and analysis of complex features, thereby enhancing the ACC of spectrum sensing. Deep learning techniques can be applied to the process of topological clustering to automatically learn network topology structures, better adapting to different network environments and topologies. Introducing elements of deep learning into the channel state information model can strengthen the modeling of relationships between nodes, improve the assessment of node trustworthiness, and reduce the risk of misjudgments caused by network uncertainties. Through deep learning-based trust modeling, the system can more effectively identify potential malicious nodes, mitigating the impact of malicious attacks on the system. Therefore, incorporating elements of deep learning into CSS not only improves the efficiency of spectrum sensing but also enhances the system's security and reliability.

CSS is a promising solution for spectrum identification. It relies on the cooperation of multiple secondary users (SU) in a CRN to detect the presence of PUs within licensed spectra. SU, through the sharing of their local sensing data and combining their observational data, can enhance the ACC and reliability of spectrum sensing. However, the collaboration in CSS is susceptible to threats, such as Byzantine attacks, where malicious nodes might provide false data to the Fusion Center (FC) or alter data during transmission, ultimately leading the system to make incorrect decisions. To address this issue, various technologies have been proposed to protect CSS from such attacks while maintaining its advantages. The following is an introduction to various techniques aimed at securing CSS:

In response to Byzantine attacks, we can adopt powerful decision fusion algorithms, such as entropy-based decision fusion, trust-based decision fusion, and cluster-based decision fusion. Deep learning provides robust tools in this field, further enhancing the system's performance and security. In entropy-based decision fusion, the application of deep learning extends to the feature extraction stage. Through deep learning models like CNNs, the system can extract more representative features from perception nodes, thereby improving the ACC of fused decisions. Deep learning can also be employed in the decision fusion process using the concept of entropy. By utilizing deep learning models to calculate the uncertainty of information, the system can intelligently allocate weights, enabling adaptive adjustments to trust levels for different decision-makers. In trust-based decision fusion, introducing deep learning, especially RNN, aids in establishing more sophisticated trust models. Through deep learning models, considerations of historical data, behavioral characteristics, and other relevant information allow for a comprehensive assessment of the credibility of decision-makers. Deep learning can also assist the system in identifying potential attackers within the trust model, thereby enhancing the system's resilience against malicious nodes such as those involved in Byzantine attacks. In cluster-based decision fusion, the concept of collaborative learning in deep learning can be integrated with cluster-based decision fusion. By utilizing deep learning and collaborative learning, nodes can engage in mutual communication and shared learning, elevating the overall system performance. This approach also helps alleviate performance degradation caused by misjudgments of individual nodes. In the decision fusion stage, deep learning models can be applied to better understand perception data, thereby enhancing collaborative decision-making. Taking into consideration the technology of deep learning, CSS systems can adapt more flexibly to different environments and improve the overall performance and security of the system in terms of decision fusion and trust modeling.

Additionally, let's introduce another method. Attack-aware Collaborative Spectrum Sensing (ACSS) estimates attack intensity by monitoring and analyzing users in the system. Attack intensity can be defined as the ratio of malicious users to the total number of users, i.e., the probability that a specific user is a malicious user. This can be estimated by analyzing user behavior, the abnormality of sensing data, communication patterns, and other factors. The K-out-N rule is a commonly used detection rule in sensing systems, where each sensing node in the system selects K nodes out of N neighboring nodes for collaborative sensing. Such a rule can be employed to enhance the system's detection performance against adversarial attacks. In the case of a given attack intensity, finding the optimal K value that minimizes Bayesian risk using Bayesian decision theory is crucial. This may involve calculating expected losses under different K values and selecting the K value that minimizes the expected loss. By comparing expected losses under different K values, the optimal K value that minimizes expected loss can be found. Deep learning can enhance the adaptive selection of the optimal K value, help establish a more accurate prior probability distribution for modeling the credibility of sensing data, and contribute to determining the most suitable collaborative sensing scale for the system given the attack intensity. Considering that attack intensity may change over time, the system should have the capability for real-time updates. Periodically or triggered by events, re-estimating attack intensity and updating K values ensures the system's adaptability and performance. Deep learning models, utilizing techniques such as online learning and transfer learning, can dynamically update the model during system runtime.

PU emulation attack is a potential security threat that may impact CSS. It primarily refers to attackers simulating or impersonating PUs to influence other nodes in the system to perform spectrum sensing, thereby affecting the effective utilization of spectrum resources. To counteract PU emulation attacks, it is necessary to implement CSS methods to enhance the security and resilience of the system.

In response to adversarial attacks, where malicious users may attempt to influence the entire system through deceptive spectrum sensing results, an adversarial-aware CSS method can be employed. This method involves analyzing the local decisions made by each CR user to identify potential adversarial behavior. Complex feature extraction and analysis can be performed using deep learning models to recognize anomalous spectrum sensing data, simulate PUs, forge channel state information, and other measures. Incorporating elements of deep learning into the design of the collaborative objective cost function allows for the flexible adaptation to different network conditions and adversarial attack patterns. By learning weights and associations, the system intelligently synthesizes information from multiple users, thereby enhancing the system's detection performance against adversarial attacks.

The application of deep learning in CSS is very important. With deep learning technology, we can more fully and accurately utilize perceptual data to improve system performance and understanding of the wireless spectrum environment. The feature extraction capability of deep learning helps the system to automatically learn key information and realize the accurate identification of spectrum features and system states. Integrated multi-source information, deep learning model improves the overall system performance and spectrum environment awareness. Deep learning makes the system more intelligent and flexible to adapt to the complex communication environment, and promotes the development of CSS technology.

2.3 Dynamic and Distributed Spectrum Access

Wireless communication has become an indispensable part of our daily lives, facilitating everything from personal mobile phone use to critical data transmission in various industries. However, with the exponential growth in demand for wireless services, wireless communication faces a significant challenge: spectrum scarcity. As mentioned in previous chapters, the spectrum refers to the range of electromagnetic frequencies used for transmitting voice, data, and video through radio waves. Spectrum is a finite resource, and its effective utilization is crucial to meet the rapidly increasing demand for wireless communication services. The scarcity of spectrum resources arises because most of the available frequencies have already been allocated, typically to broadcasting services, military applications, and mobile network operators. Traditionally, this allocation has been static, with specific bands dedicated to specific uses, often leading to inefficient utilization of these bands. With the development of wireless technology and increasing demand for bandwidth, this static allocation model has proven to be inefficient, resulting in underutilization of spectrum in some areas while causing bottlenecks in others.

To address this challenge, DSA has emerged as a promising solution. Dynamic SA is a set of techniques and technologies that allow for more flexible access to the spectrum. Instead of fixed allocations, DSA enables wireless devices to dynamically access spectrum bands on an as-needed basis. This means that unused frequencies can be utilized by other devices, significantly improving the efficiency of spectrum use. DSA relies on sophisticated algorithms and sensing technologies to detect which frequencies are not being used at a particular time and location and then temporarily allocates those frequencies to users or devices that need them. Distributed spectrum access is an extension of the DSA concept, focusing on decentralizing the decision-making process regarding spectrum access. In a distributed spectrum access system, individual devices or networks make decisions about spectrum use based on local observations and negotiations with neighboring devices, rather than relying on a central authority. This approach can further enhance spectrum efficiency by adapting to local conditions in real-time and reducing the overhead and latency associated with centralized control.

The importance of DSA and distributed spectrum access cannot be overstated. They offer a viable solution to the spectrum scarcity problem by maximizing the utilization of available frequencies. This enhanced efficiency can lead to increased capacity, better quality of service (QoS), and support for a greater number of users and devices. Furthermore, these approaches can foster innovation in wireless services and applications, stimulate economic growth, and help meet the ever-growing demand for wireless communication. However, the implementation of these strategies is not without its challenges. Adapting to dynamic environments is a cornerstone of both DSA and distributed spectrum access, necessitating devices and networks to continuously monitor the spectrum landscape to identify unused frequencies. This requirement introduces the challenge of ensuring spectrum sensing ACC, where the goal is to detect available channels accurately without causing interference to incumbent users. The task is made more complex by the need to respond swiftly to rapid changes in spectrum availability, demanding advanced technologies that can keep pace with such fluctuations. Interference management among users is another significant hurdle. As these strategies aim to enable more users to access the spectrum, it becomes imperative to develop mechanisms that allow for the harmonious coexistence of multiple users. This involves not only technical solutions to prevent harmful interference but also policy frameworks that clearly define the rights and priorities of primary (licensed) and secondary (unlicensed) users. Striking a balance between protecting the interests of incumbent users and maximizing spectrum efficiency presents a nuanced challenge. Moreover, decision-making delays pose a critical concern in the real-time allocation of spectrum resources. The process of collecting and processing data on spectrum usage, coupled with the need for decentralized coordination among multiple devices, can introduce delays that compromise the effectiveness of spectrum allocation decisions. Additionally, ensuring compliance with regulatory requirements further complicates the decision-making process, potentially slowing down the dynamic allocation of spectrum resources.

In conclusion, while DSA and distributed spectrum access present innovative solutions to the pressing issue of spectrum scarcity in wireless communication, they also introduce a set of complex challenges. These include the need for rapid adaptation to dynamic environments, effective interference management among users, and the minimization of decision-making delays. Addressing these challenges is crucial for unlocking the full potential of these spectrum access strategies. In this context, DRL emerges as a powerful tool with significant potential to overcome these obstacles. DRL, a subset of AI, is adept at making decisions in complex, dynamic environments by learning optimal actions through trial and error. Its ability to adapt and learn from the environment makes it particularly well-suited for applications in DSA and distributed spectrum access, where conditions constantly change and swift, autonomous decision-making is essential. The application of DRL can lead to more efficient spectrum sensing techniques, enhance the capability of systems to manage interference among a multitude of users, and significantly reduce decision-making delays by enabling real-time, intelligent decision-making processes. As such, DRL holds the promise of not only addressing the challenges inherent in DSA and distributed spectrum access but also of propelling these strategies towards their full efficacy and potential. Thus, the integration of DRL into DSA and distributed spectrum access represents a promising frontier in the quest for more efficient and flexible wireless communication systems. By harnessing the power of DRL, we can navigate the complexities of DSA, paving the way for a future where spectrum scarcity is effectively mitigated through intelligent, adaptive technologies.

In DSA, DRL offers targeted solutions in terms of spectrum sensing. This process is crucial for enabling wireless devices to effectively discover and utilize unused frequencies without interfering with existing users.

The working principle of DRL involves an agent learning to make decisions through its interactions with the environment. In DSA, the "agent" could be a CR or a network device equipped with DRL capabilities, and the "environment" refers to the spectrum landscape, which is continuously changing due to varying usage patterns. The primary goal of a DRL agent is to identify parts of the spectrum that are temporarily unoccupied and can be used for data transmission. The process begins with the DRL agent observing the state of the spectrum, including collecting data on various frequencies to determine their current usage status. Based on this observation, the agent takes an action, such as selecting a specific band for transmission. This action is then evaluated through feedback (usually in the form of rewards or penalties). For instance, if the selected band is indeed free and the transmission is successful without causing interference, the agent receives a positive reward. Conversely, if the transmission interferes with existing users, it is penalized. Over time, through a process of trial and error, the DRL agent learns to predict which actions (i.e., selecting specific bands) will maximize its rewards - essentially learning to identify available spectrum with high precision. This learning process is facilitated by sophisticated algorithms that enable the agent to adjust its strategy based on past experiences and emerging patterns in spectrum usage.

One of the key advantages of using DRL for spectrum sensing in DSA is its ability to adapt to complex and dynamic environments. Unlike traditional methods that may rely on predefined rules or static databases, DRL continuously evolves its strategy, improving its ability to detect available spectrum as the environment changes. This capability is particularly crucial in densely populated areas or during peak usage times when the availability of spectrum can fluctuate rapidly. Furthermore, DRL can help in efficiently managing the trade-offs between exploring new frequency bands and exploiting known unoccupied bands. This balance is critical for optimizing spectrum utilization while minimizing the risk of interference with incumbent users.

In summary, DRL presents a robust and adaptive approach for enhancing spectrum sensing in DSA. By leveraging DRL, wireless devices can more effectively identify and utilize available spectrum, thereby addressing a critical challenge in maximizing the efficiency of spectrum usage. Through its ability to learn and adapt in real-time, DRL holds significant promise for advancing the capabilities of DSA systems, paving the way for more flexible and efficient wireless communication networks.

DSA has emerged as a pivotal strategy in addressing the challenge of spectrum congestion by allowing wireless devices to dynamically utilize underutilized spectrum bands. This approach promises to enhance spectrum efficiency significantly; however, it introduces the complex task of making real-time, adaptive decisions regarding frequency band selection and power level adjustments. In this context, DRL offers a powerful solution by enabling systems to learn optimal policies through continuous interaction with their environment.

The essence of DRL lies in its ability to integrate perception, decision-making, and action into a cohesive framework. Here, an agent learns to make informed decisions by engaging with an environment, aiming to maximize a cumulative reward over time. Applied to DSA, the environment in question is the wireless spectrum, with actions encompassing the choices around which frequency bands to access and the determination of appropriate power levels for transmission. The state in a DRL-based DSA system encapsulates information about current spectrum usage, channel quality, and historical occupancy data, alongside the agent's current power level and past actions. Such a comprehensive state representation is crucial for making informed decisions. The action space in DRL for DSA includes potential decisions the agent can make, such as selecting frequency bands for access and determining suitable power levels for transmission. These actions are critical for optimizing spectrum usage and ensuring efficient communication. Furthermore, the reward function in DRL is meticulously designed to steer the agent towards optimal behavior, incorporating factors like successful transmission rate, interference minimization, and energy efficiency. Through this, the agent is encouraged to select less congested frequency bands and adjust power levels to balance interference and throughput optimally. DRL empowers wireless devices with several strategies for effective spectrum allocation. It enables agents to sense the spectrum environment accurately and select the best available frequency bands based on real-time dynamics. This adaptability is a significant leap over traditional methods, which may not effectively predict or adapt to changing spectrum usage. Additionally, DRL agents can dynamically adjust their transmission power levels to maintain communication quality while minimizing interference, offering a level of control that rule-based approaches struggle to achieve. Moreover, in scenarios involving multiple DRL agents, such as different network operators, these agents can learn to coordinate their actions, optimizing overall spectrum utilization without the need for centralized control mechanisms.

Compared to traditional spectrum allocation methods, DRL offers substantial advantages. Its ability to adapt to dynamic and uncertain environments aligns well with the variable nature of wireless spectrum usage, surpassing static rule-based or model-dependent traditional methods. DRL also excels in optimizing complex objectives, balancing throughput, interference, and energy efficiency in a way that traditional methods, often focused on single objectives, cannot. Furthermore, DRL's support for decentralized decision-making enhances scalability and flexibility in managing spectrum resources across various users and devices, presenting a significant improvement over centralized control systems.

DRL provides a sophisticated and adaptable approach to DSA, enabling more efficient and intelligent spectrum allocation decisions. In the intricate landscape of DSA, managing and sharing spectrum resources in multi-user scenarios poses a formidable challenge. This complexity is rooted in the necessity to cater to diverse demands, mitigate interference, and ensure equitable access for a multitude of users. Emerging as a potent solution to navigate these complexities, Multi-Agent Reinforcement Learning (MARL) has demonstrated its capability to enable effective spectrum sharing and foster collaboration among users. By deploying multiple learning agents that interact within the shared environment of the RF spectrum, MARL transcends traditional approaches by learning optimal policies for spectrum access through the lens of both individual needs and collective system efficiency.

At the heart of MARL's approach to DSA is the principle of decentralized decision-making. Unlike centralized control schemes, where a singular entity dictates the allocation of spectrum resources, MARL empowers each agent to make informed decisions based on local observations and, crucially, on global communication cues. This decentralization not only enhances scalability, allowing the system to accommodate an increasing number of users seamlessly but also ensures that decisions are made with a comprehensive understanding of the current spectrum environment. Agents, through their interactions, learn to select frequency bands, adjust transmission powers, or vacate bands to minimize interference, all while maximizing the utility of the spectrum. Furthermore, MARL facilitates collaboration among agents by developing cooperative strategies that take into account the actions and intentions of others. This aspect is vital in DSA, where the goal extends beyond individual optimization to encompass the efficient use of spectrum without causing detrimental interference to others. Through feedback mechanisms and repeated interactions, agents learn to predict the behavior of their counterparts and adapt their strategies accordingly. Such collaborative efforts can lead to dynamic alternation in accessing specific frequency bands, ensuring fair transmission opportunities for all users involved.

The advantages of MARL over traditional spectrum management methods are manifold. Firstly, the adaptability inherent in MARL systems allows for real-time responses to changes in the spectrum environment, user behaviors, or regulatory policies, a feat unachievable by static allocation methods. Secondly, MARL's focus on optimizing the actions of individual agents for the collective benefit significantly improves spectrum utilization, ensuring that available bands are used efficiently. Thirdly, the scalability offered by the decentralized nature of MARL means that the system can easily integrate new users without significant disruption. Lastly, MARL algorithms can be designed with fairness in mind, distributing resources equitably among users and overcoming the limitations of first-come, first-served or fixed allocation schemes.

In essence, MARL heralds a new era in the management of DSA, particularly in environments populated by multiple users. Its capacity for enabling effective spectrum sharing and fostering collaboration, coupled with its advantages in adaptability, efficiency, scalability, and fairness, positions MARL as a superior alternative to conventional spectrum management techniques. As the demand for wireless communication services continues to soar, the integration of MARL into DSA frameworks promises to unlock unprecedented levels of spectrum efficiency, navigating the complexities of the modern wireless landscape with sophistication and dynamism.

DRL is revolutionizing the way we approach DSA in distributed spectrum access systems, presenting a paradigm shift from traditional methods towards a more adaptive, efficient, and scalable framework. At the heart of this transformation is the concept of strategy learning, where DRL agents interact with the wireless environment, making informed decisions on spectrum access that aim to maximize a cumulative reward. These agents embark on a journey of exploration and exploitation, navigating through various actions to unearth strategies that yield the most favorable outcomes. As they progress, the agents develop a nuanced understanding of which actions to take in specific environmental states, crafting an optimal policy for accessing the spectrum.

The process is underpinned by sophisticated environmental modeling, a crucial step that encapsulates the dynamics of the spectrum environment. This modeling allows DRL agents to predict the impact of their actions and to anticipate future environmental states, facilitating more informed decision-making. Techniques such as NNs play a pivotal role here, offering a means to approximate the intricate relationships between actions, states, and rewards. Moreover, this modeling extends to account for the interactions between multiple users within the spectrum, a critical consideration in distributed systems where users operate independently. By accurately capturing how one user's actions affect the spectrum availability for others, DRL agents can learn to adopt cooperative strategies that optimize spectrum usage while minimizing conflicts. What sets DRL apart from traditional spectrum management methods are its inherent advantages in adaptability, efficiency, scalability, and fairness. Unlike static allocation schemes that struggle to keep pace with the dynamic nature of the spectrum environment, DRL agents continuously refine their strategies based on ongoing interactions with the environment. This dynamic adaptation ensures that spectrum resources are utilized efficiently, even as conditions change. Furthermore, the decentralized approach of DRL aligns perfectly with the ethos of distributed spectrum access systems, allowing for seamless scalability as new users join the system without necessitating centralized coordination. Additionally, DRL algorithms can be tailored to incorporate fairness, ensuring that all users have equitable access to spectrum resources, thereby balancing individual and collective objectives.

DRL presents a powerful approach for achieving DSA in distributed spectrum access systems. Through sophisticated strategy learning and environmental modeling, DRL enables more adaptable, efficient, and scalable spectrum management, offering substantial improvements over traditional methods.

In the realm of distributed spectrum access, orchestrating the allocation of the limited RF spectrum among myriad users necessitates a sophisticated approach to ensure optimal sharing and minimize interference. The advent of MARL has ushered in a promising strategy to tackle this intricate challenge. MARL enables multiple agents, each representing different users or devices, to learn and adapt their strategies for accessing the spectrum in a dynamic and decentralized environment. This learning paradigm hinges on the principle of agents interacting with their surroundings and each other, striving to maximize their cumulative rewards through their decision-making processes.

The application of MARL in distributed spectrum access is characterized by decentralized decision-making, where each agent operates independently based on local observations. This autonomy not only fosters scalability but also imbues the system with flexibility. Furthermore, the framework facilitates coordination among agents through mechanisms such as policy sharing and joint action learning. Agents can share insights or learned policies with their peers, paving the way for cooperative strategies that enhance spectrum access and reduce interference. Additionally, by predicting the actions of other agents, individuals can synchronize their strategies, thereby curtailing collisions and optimizing resource utilization. A pivotal advantage of MARL lies in its ability to dynamically adapt to changing network conditions and user demands, outshining traditional fixed allocation schemes. Agents learn to identify underutilized spectrum bands and adjust their access patterns accordingly, thus improving overall spectrum utilization. Moreover, by understanding the patterns of interference among agents, MARL empowers them to alter their strategies to minimize harmful interference, ensuring a higher QoS. The design of reward structures plays a crucial role in guiding agents to strike a balance between maximizing spectrum access and minimizing interference, encapsulating the essence of optimized shared resource management.

Compared to conventional methods, MARL offers significant benefits, including dynamic adaptation to fluctuating environments, scalability to accommodate a large number of users without centralized control, and enhanced robustness against failures and environmental changes. Its suitability for complex scenarios, where interactions among users and between users and the environment are unpredictable, underscores its efficiency and potential to revolutionize distributed spectrum access.

MARL presents a compelling approach for coordinating distributed spectrum access, offering significant advantages in terms of adaptability, scalability, and efficiency. By enabling dynamic, decentralized decision-making and optimizing the shared use of spectrum while managing interference, MARL can significantly outperform traditional methods in managing the complex challenges of distributed spectrum access. As the technology and methodologies around MARL continue to evolve, its application in spectrum access is poised to facilitate more efficient and flexible wireless communication systems.

The integration of DRL into DSA signifies a forward leap in optimizing the utilization of spectrum resources. DSA's adaptive mechanism, which allows wireless devices to modify their transmission parameters to minimize interference, pairs well with DRL's ability to learn and make decisions from complex, dynamic environments. However, this promising amalgamation is not devoid of challenges, particularly concerning the stability of learning algorithms and the efficiency of the training process.

Ensuring algorithm stability in DRL applications within DSA is a critical hurdle. The unpredictable nature of the spectrum environment, characterized by fluctuating availability, user demand, and interference patterns, introduces high variance in learning rewards. This variance complicates the convergence of DRL agents to a stable policy, further complicated by the non-stationary environment where other users' adaptive strategies create a continuously evolving target for the DRL agent. Techniques such as experience replay, target networks, and reward shaping have been developed to mitigate these issues, aiming to stabilize the learning process. Despite these advancements, balancing exploration with exploitation remains a significant challenge in the volatile DSA landscape. Training efficiency poses another significant challenge. DRL models, especially those based on DNNs, demand extensive computational resources and vast amounts of interaction data to derive effective policies. Given the complexity and dynamism of the DSA environment, the volume of data required for effective training is substantial, raising concerns for real-world applicability where computational resources are finite, and swift decision-making is essential. The high-dimensional state and action spaces of DSA exacerbate the training difficulty, leading to prolonged training durations and delayed model convergence. Researchers have explored various strategies to enhance training efficiency, including transfer learning, curriculum learning, and parallel computing techniques. These efforts aim to streamline the training process in the demanding DSA context, yet efficiently training DRL models remains an arduous task.

At the same time, DRL offers a novel approach to tackling the intricate issue of distributed spectrum access in wireless communication networks, promising to optimize network performance and minimize interference among users. However, the deployment of DRL in this domain is fraught with challenges, notably concerning algorithm stability, convergence speed, and sample efficiency, each of which plays a pivotal role in the successful application of this technology.

The stability of DRL algorithms is paramount, yet difficult to ensure in the volatile environment of distributed spectrum access. The rapid changes in spectrum state, driven by user mobility, varying traffic loads, and fluctuating signal strengths, create a non-stationary environment that can destabilize the learning process. This instability is further compounded by the exploration-exploitation dilemma inherent in RL, where aggressive exploration strategies can introduce high variance in observed rewards, complicating the convergence to a stable policy. Convergence speed represents another significant hurdle. The vast state and action spaces characteristic of wireless networks slow the learning process, demanding extensive computation and time. This challenge is exacerbated by the necessity for coordination among multiple agents, whose policies not only impact the overall network performance but also influence each other's learning outcomes. Such interdependencies can induce oscillations in policy updates, impeding swift convergence. Moreover, achieving high sample efficiency in distributed spectrum access is an arduous task due to the environment's complexity and the delayed, sparse rewards. Agents must navigate through a sequence of interactions before the consequences of their spectrum access decisions become apparent, making it challenging to attribute outcomes to specific actions. This requirement for a large number of samples to accurately estimate action values is intensified by the dynamic actions of other agents, which add to the non-stationarity of the environment and demand even more extensive sampling for effective policy adaptation.

In summary, the application of DRL in DSA holds immense potential to revolutionize wireless communication networks by enabling intelligent and dynamic spectrum management. Despite the significant challenges related to algorithm stability, convergence speed, and sample efficiency, ongoing research and technological advancements promise to address these hurdles effectively. As we move forward, we can anticipate the development of more sophisticated DRL algorithms that are not only robust and adaptive but also capable of facilitating efficient coordination among multiple agents in highly dynamic environments. These advancements are expected to enhance the overall performance of wireless networks, ensuring optimal spectrum utilization and significantly reducing interference among users. The future of DRL in DSA and distributed spectrum access appears promising, with the potential to usher in a new era of intelligent wireless communication systems that are more resilient, efficient, and capable of meeting the ever-growing demands for connectivity in our digital world.

2.4 Supervised Learning-Based Spectrum Sharing

Wireless communication systems rely on the efficient utilization of limited frequency spectrum resources to provide reliable and high-speed data transmission. Frequency spectrum sharing is a key concept in wireless communication, which refers to the practice of allowing multiple users or services to share the same frequency bands without causing interference. Frequency spectrum, in simple terms, can be thought of as a range of radio frequencies over which wireless signals can be transmitted. However, the available frequency spectrum is finite, and different wireless services require specific frequency bands to operate effectively. This is where frequency spectrum sharing becomes crucial.

To enable efficient spectrum sharing, regulatory authorities allocate different frequency bands for various wireless services such as cellular networks, Wi-Fi, bluetooth, satellite communications, and television broadcasting. These allocations are based on international agreements and national regulations to ensure organized and interference-free communication. One of the most widely used techniques for frequency spectrum sharing is known as "licensed spectrum sharing". In this approach, certain frequency bands are exclusively assigned to specific users or services through licensing. For example, cellular network operators obtain licenses to use particular frequency bands for their network infrastructure. This ensures that they have dedicated spectrum resources and can provide reliable services without interference from other users. Another technique for spectrum sharing is called 'unlicensed spectrum sharing,' where certain frequency bands are made available for public use without requiring a license. This approach fosters accessibility but necessitates adherence to specific rules and regulations for fair and efficient sharing. In this case, certain frequency bands are made available for public use without requiring a license. The best-known example is the unlicensed spectrum used by Wi-Fi networks. These frequency bands are open for anyone to access, but users must adhere to specific rules and regulations to ensure fair and efficient sharing.

However traditional spectrum sharing methods, such as exclusive licensing and static allocation, face a range of challenges in meeting the growing demand for wireless communication services. One of the primary issues is the increasing need for spectrum resources driven by the proliferation of mobile devices, IoT applications, and emerging wireless technologies like 5G and beyond. This surge in demand has led to spectrum scarcity, making it difficult to allocate sufficient bandwidth to support the expanding array of services. Secondly, the inefficient utilization of available spectrum bands poses a significant challenge. The licensed spectrum remains underutilized or unused for extended periods due to static and rigid allocation policies, limiting opportunities for dynamic sharing among multiple users and services. This means that some frequency bands remain unused while others may suffer from congestion. As a result, there is a lack of efficient spectrum utilization. At the same time, traditional methods rely on centralized control and coordination, which can lead to inefficiencies and delays in spectrum access. This central authority must carefully manage and allocate frequencies, which may result in long wait times for users and limited flexibility in adapting to dynamic changes in demand. This inefficiency is exacerbated by the lack of flexibility in spectrum allocation policies, which often do not adapt well to changing usage patterns and service requirements. Furthermore, interference among different users and services is a significant challenge in traditional spectrum sharing. When multiple users attempt to access the same frequency band simultaneously, interference can occur, degrading the QoS for all users involved. Lastly, traditional spectrum sharing methods often lack adaptability and scalability. They are designed for specific frequency bands and technologies, making it difficult to accommodate new and emerging technologies. The static and exclusive nature of traditional spectrum allocation models hinders effective spectrum sharing. Fixed spectrum assignments to specific license holders restrict the ability to dynamically reallocate resources based on varying demand levels or to accommodate new entrants or temporary users. This rigidity can result in underutilization of spectrum in some areas while leading to congestion and interference in others. This restricts innovation and hampers the development of wireless communication systems.

In contrast, DRL presents promising potential for spectrum sharing. By leveraging artificial intelligence techniques, DRL enables dynamic and autonomous decision-making in spectrum access, leading to more efficient and flexible spectrum utilization. Furthermore, it facilitates decentralized decision-making, reducing the need for centralized control and coordination, thereby enhancing scalability and responsiveness in spectrum sharing. The utilization of DRL in conjunction with supervised learning-based spectrum sharing offers several distinct advantages over traditional spectrum sharing methods. Traditional spectrum sharing approaches often rely on static allocation policies that may not effectively adapt to dynamic and complex wireless environments. In contrast, the integration of DRL and supervised learning enables a more adaptive and intelligent spectrum sharing strategy. DRL empowers systems to learn optimal spectrum access policies through interaction with the environment and feedback mechanisms. This dynamic learning process allows for real-time adjustments and improvements in spectrum allocation decisions, enhancing system efficiency and adaptability. Supervised learning plays a critical role in providing accurate and reliable spectrum sensing and classification capabilities. By leveraging labeled training data, supervised learning algorithms can effectively detect PUs and identify available spectrum opportunities, thereby minimizing interference and maximizing spectrum utilization. By combining the strengths of DRL and supervised learning, the spectrum sharing system can make informed decisions based on contextual information and real-time feedback. This approach leads to enhanced spectral efficiency, reduced interference, and improved coexistence among multiple users sharing the same frequency bands. Overall, the application of DRL in supervised learning-based spectrum sharing represents a significant advancement over traditional methods. It enables more intelligent, adaptive, and efficient spectrum management, ultimately leading to better utilization of available spectrum resources and improved performance in dynam.

Supervised learning is a machine learning technique that has found applications in various fields, including spectrum sensing and spectrum allocation. In these domains, supervised learning plays a crucial role in guiding spectrum sharing decisions based on historical data and labeled samples. In spectrum sensing, the goal is to detect the presence or absence of signals in a given frequency band. Supervised learning can be employed to train models that can accurately classify different types of signals, such as licensed and unlicensed users, primary and SU, or different modulation schemes. This involves collecting a dataset of historical spectrum measurements along with their corresponding labels, indicating the presence or absence of specific signals. By using this dataset, a supervised learning algorithm can learn patterns and characteristics of different signals, enabling accurate detection and classification of signals in real-time scenarios. Furthermore, supervised learning can also be applied to spectrum allocation, which involves determining how to assign available spectrum resources to different users or systems. By utilizing historical data on spectrum usage and performance metrics, supervised learning algorithms can learn the relationship between spectrum allocation decisions and their impact on system performance, such as throughput, interference, or energy efficiency. This enables the algorithm to make informed decisions regarding optimal spectrum allocation strategies, thus maximizing the utilization of spectrum resources. In traditional spectrum sharing methods, where spectrum is allocated based on predetermined rules or fixed policies, supervised learning introduces a data-driven approach. By leveraging historical data and labeled samples, supervised learning allows for more adaptive and dynamic spectrum sharing decisions. Instead of relying solely on predefined rules, the algorithm can learn from past experiences and adjust its decisions based on changing environmental conditions or user requirements. This flexibility leads to improved efficiency and fairness in spectrum allocation. It is worth noting that supervised learning in spectrum sensing and allocation heavily relies on the availability of high-quality labeled data. The process of labeling the data often requires domain expertise and careful annotation, ensuring the ACC and reliability of the training process. Additionally, continuous updates and retraining of the models are necessary to adapt to evolving spectrum dynamics and user behaviors.

Combining DRL with supervised learning offers a powerful approach that leverages the strengths of both techniques. In the context of spectrum sharing, this fusion can lead to more effective and efficient decision-making processes. DRL excels in learning optimal decision-making policies through interaction with the environment and receiving rewards based on its actions. By combining DRL with supervised learning, we can enhance the learning process by incorporating labeled data and historical information to guide the decision-making of the agent. In the realm of spectrum sharing, the synergy between DRL and supervised learning is particularly beneficial. DRL can learn complex strategies for spectrum allocation and access by interacting with the dynamic spectrum environment. On the other hand, supervised learning can provide valuable insights based on past data and labeled samples, helping the agent make more informed decisions. For example, in spectrum sharing scenarios, DRL can continuously explore different spectrum allocation strategies and adapt based on feedback from the environment. Meanwhile, supervised learning can contribute by offering guidance based on historical performance data, user behavior patterns, and regulatory constraints. This combined approach enables the agent to learn efficiently from both its interactions with the environment and the knowledge distilled from past experiences.

Moreover, the integration of DRL and supervised learning in spectrum sharing can lead to improved spectrum utilization, reduced interference, and enhanced overall system performance. The agent can leverage the strengths of each technique to make intelligent decisions that balance the trade-offs between different users' needs and maximize the efficiency of spectrum usage. For instance, in the realm of wireless communication, the fusion of supervised learning-based spectrum sensing with DRL-based spectrum allocation presents a powerful approach for optimizing spectrum utilization. Firstly, in supervised learning-based spectrum sensing, algorithms are trained to accurately detect and classify the presence of PUs in specific frequency bands. By leveraging labeled training data, these algorithms can effectively identify available spectrum opportunities for SU without causing harmful interference. On the other hand, DRL-based spectrum allocation focuses on making dynamic decisions regarding how to allocate available spectrum resources among competing users. Through continuous interaction with the environment, an agent learns optimal strategies to maximize overall network performance while adhering to specific constraints and objectives. When these two methodologies are combined, the system gains significant advantages over using either approach individually. The supervised learning component enhances spectrum sensing ACC, providing reliable input to the RL agent for more informed decision-making. Meanwhile, the RL aspect enables adaptive and efficient spectrum allocation strategies based on real-time feedback, leading to improved overall system efficiency and user satisfaction. By integrating supervised learning-based spectrum sensing with DRL-based spectrum allocation, this fusion method offers a comprehensive and sophisticated solution for enhancing spectrum management in dynamic and complex wireless environments.

The future prospects of applying DRL to supervised learning-based spectrum sharing are highly promising. As wireless communication systems continue to evolve towards more dynamic, heterogeneous, and densely populated environments, the need for intelligent and adaptive spectrum sharing mechanisms becomes increasingly critical. DRL offers the potential to revolutionize spectrum sharing by enabling systems to learn and adapt their spectrum access strategies based on real-time feedback and environmental changes. This adaptability is particularly well-suited for addressing the challenges associated with dynamic and unpredictable wireless environments, where traditional static allocation schemes may prove insufficient. Furthermore, the combination of DRL with supervised learning-based spectrum sharing can lead to more accurate and efficient spectrum sensing and classification, allowing for better utilization of available spectrum resources. This, in turn, can reduce interference and enhance overall spectrum efficiency. Looking ahead, the application of in supervised learning-based spectrum sharing holds the promise of enabling more intelligent, context-aware, and adaptive spectrum access, which is essential for supporting the growing demands of diverse wireless services and applications.

Chapter 3
Learning Resource Allocation Optimization

3.1 Resource Allocation with Unsupervised Learning

Transmission rates and the popularity of wireless communication devices, especially with the introduction of 5G technology, are experiencing exponential growth. This creates problems with limited spectrum resources and growing communication needs. In this context, unsupervised learning is applied to resource allocation to better meet the actual needs and to uncover the hidden resource utilization patterns and rules. Especially in optimizing continuous power control, unsupervised learning methods are more efficient.

In traditional resource allocation problems, iterative algorithms are often used. However, the solutions obtained by these iterative algorithms are usually suboptimal. When the number of users or the number of resource allocation variables is large, iterative methods may take a long time to converge, limiting their application in real-time operations. In addition, resource allocation solutions based on optimization are specific solutions to specific problems, and if the form of the problem changes, completely new solutions must be developed, further limiting the flexibility of the application. Deep learning approaches can overcome these limitations.

At present, the application of resource allocation schemes based on deep learning in channel estimation and data detection of orthogonal frequency division multiplexing (OFDM) and filter bank multi-carrier modulation has been studied. In addition, the channel state information feedback, encoder and decoder design of millimeter wave multiple-input multiple-output (MIMO) system based on deep learning are studied, and the autoencoder based on deep neural network (DNN) is applied for optical wireless communication. These studies verify the practical relevance and feasibility of deep learning-based wireless communication system design in solving resource allocation problems.

In a deep learning-based resource allocation scheme, first, it is able to achieve the best performance without solving complex optimization problems. DNN can simulate the optimal solution of resource allocation problem, so as to improve the

H. Zhang and N. Yang, *Deep Learning in Wireless Communications*,
https://doi.org/10.1007/978-981-97-6314-6_3

efficiency of resource allocation. In addition, by training the model, it is possible to find strategies to solve resource allocation problems more quickly. Second, deep learning-based resource allocation schemes are more flexible and can achieve different design goals by changing the loss function and training the DNN structure. Third, because the trained DNNS only perform simple matrix operations, the computation time required to obtain resource allocation strategies is much less than that of traditional schemes.

To sum up, deep learning has an important application prospect in the resource allocation of wireless communication systems. Resource allocation schemes based on unsupervised learning do not require labeled data, and DNN can autonomously determine the best resource allocation strategy. Compared with the scheme based on supervised learning, the scheme based on unsupervised learning is simpler and easier to implement. For challenging scenarios where the environmental model is known or at least the mathematical model is known but the environmental state distribution is unknown, we can employ an unsupervised algorithm to learn the optimal solution to the environmental state. In addition, unsupervised learning can provide better performance when the global channel state information is limited and the local channel state information sharing strategy is not optimized.

To utilize unsupervised learning for solving resource allocation problems in cellular networks, we can employ a DNN approach. Here's a step-by-step process:

1. Define The Problem: Clearly outline the resource allocation problem, taking into account factors like channel bandwidth W, the number of channels M, the transmit power of the PU p_o^m , the transmit power of SU, noise spectral density No, and the channel gain $h_{i,j}^m$ between transmitters and receivers of SUs.

2. Normalize Inputs: Normalize the inputs to improve the learning process. For example, use a normalization formula

$$\hat{h_{i,j}^m} = \frac{\log_{10} h_{i,j}^m - \mu}{\delta}, \mu = \mathcal{E}\left[\log_{10} h_{i,j}^m\right], \delta = \sqrt{\mathcal{E}\left[\log_{10} h_{i,j}^m - \mu\right]^2} \quad (3.1)$$

to normalize the channel gain which is $h_{i,j}^m$ and transmit power which is p_o^m of the PU.

3. Design the DNN Architecture: Construct a DNN with an input layer, several hidden layers, and an output layer. Each hidden layer should incorporate the ReLU activation function, with P series of fully connected hidden layers, each containing Q neurons. The output layer should employ a softmax function.

4. Prepare Training Data: Gather training data that includes input feature vectors (normalized channel gain and transmit power) and corresponding output labels (desired resource allocation decisions). In this case, the output labels would represent the realizable rate of each SU pair

$$R_{\rangle}(\mathbf{p}) = \sum_{m \in M} W \log_2 \left(1 + \frac{h_{i,i}^m p_i^m}{N_o W + \sum_{l \in N \backslash \{i\}} h_{i,i}^m p_l^m + h_{o,i}^m p_o^m}\right) \quad (3.2)$$

5. Train the DNN: Utilize an unsupervised learning algorithm like backpropagation to train the DNN on the collected training data. The DNN learns the underlying patterns and features that optimize the sum rate of the cognitive radio network (CRN).

6. Obtain Solution: Once the DNN is trained, the output of the network represents the resource allocation decisions. Multiply the network output by a constant $\sum_{m \in M} p_o^m$ to acquire the solution $\mathbf{p}$ to the optimization problem, as outlined in the problem statement. In order to meet the constraints in the problem, we can constantly update the parameters of DNN to maximize $\sum_{i \in N} R_i(\mathbf{p})$ by training the loss function

$$\begin{aligned} L = & -\lambda_1 \sum_{i \in N} R_i(\mathbf{p}) + \lambda_2 \sum_{m \in M} \tanh\left(\frac{\left[\sum_{i \in N} h_{i,o}^m p_i^m - I_{thr}\right]^+}{I_{thr}}\right) \\ & + \lambda_3 \sum_{i \in N} \tanh\left(\frac{\left[R_{thr} - R_i(\mathbf{p})\right]^+}{R_{thr}}\right) N \end{aligned} \tag{3.3}$$

λ_1,λ_2 and λ_3 has a great impact on the training results. When a certain weight is too large, the QoS of PU and SU may not be guaranteed. On the contrary, the training results should ensure the QoS of PU and SU as much as possible. The simulation results show that the scheme can significantly improve the total sum rate of SUs, and the calculation time is short.

7. Evaluate and Refine: Assess the performance of the DNN's resource allocation decisions using validation or test data. If necessary, refine the DNN architecture, training process, or input features to enhance its performance.

By following these steps, unsupervised learning with a DNN can be employed to learn optimal resource allocation decisions based on the input features, thereby improving the sum rate of the cellular network.

3.2 DRL for Resource Allocation

Resource allocation plays a crucial role in various domains, such as computer networks, cloud computing, and the IoT. Efficient resource allocation ensures optimal utilization of available resources and contributes to the overall performance and effectiveness of these systems.

In computer networks, resource allocation involves distributing network bandwidth, processing power, and storage capacity among different applications, devices, or users. Effective resource allocation is essential for ensuring smooth and uninterrupted communication, minimizing latency, and maximizing network throughput. By allocating resources based on demand and priority, computer networks can handle varying workloads and prioritize critical tasks, providing a seamless and responsive user experience. In cloud computing, resource allocation is vital for delivering scalable and on-demand services to customers. Cloud service providers must efficiently allocate computing resources, such as virtual machines, storage, and network bandwidth, to meet customer demands while optimizing resource utilization and minimizing costs. Effective resource allocation enables dynamic scaling, allowing customers to scale their resources up or down based on their needs, ensuring efficient use of cloud infrastructure and enhancing the overall performance and cost-effectiveness of cloud-based applications and services. The IoT connects a vast number of devices and sensors, generating massive amounts of data. Resource allocation in IoT involves managing limited resources, such as battery power and network bandwidth, among numerous interconnected devices. Efficient resource allocation is crucial for prolonging device battery life, optimizing data transmission, and ensuring reliable and timely communication between devices. By allocating resources intelligently, IoT systems can enhance energy efficiency, reduce communication overhead, and improve overall system reliability and responsiveness.

In addition to these specific domains, resource allocation is also important in many other areas, such as transportation logistics, energy management, and manufacturing. In transportation logistics, resource allocation involves optimizing routes, scheduling vehicles, and assigning tasks to minimize costs and maximize efficiency. In energy management, resource allocation focuses on balancing energy supply and demand, optimizing energy generation and distribution, and promoting renewable energy utilization. In manufacturing, resource allocation aims to optimize production processes, assign tasks to machines or workers, and minimize production costs while ensuring timely delivery and high-quality products.

Resource allocation is of utmost importance in various domains. It ensures efficient utilization of available resources, enhances system performance, and contributes to cost-effectiveness and user satisfaction. Whether it is in computer networks, cloud computing, IoT, or other areas, effective resource allocation plays a pivotal role in optimizing system operations and achieving desired outcomes.

Traditional resource allocation methods often face limitations in addressing the complexity and dynamics of modern distributed systems. These methods, which may rely on static rules, predefined heuristics, or centralized control, encounter challenges in adapting to changing environments, optimizing resource utilization, and achieving scalable and efficient allocations. One major limitation of traditional resource allocation methods is their difficulty in handling dynamic and uncertain environments. Systems with fluctuating workloads, varying resource demands, and evolving conditions present a challenge for static allocation approaches, which struggle to adapt and reconfigure resources in real time. Additionally, traditional methods may lack the ability to consider the global system state and make coordinated decisions across distributed nodes, leading to suboptimal resource allocations and potential inefficiencies. Moreover, traditional resource allocation approaches often rely on centralized control or decision-making, which can introduce single points of failure, scalability issues, and communication overhead. Centralized systems may struggle to scale effectively as the number of resources and participants grows, leading to bottlenecks and reduced responsiveness. Furthermore, centralized control may not fully leverage the local knowledge and decision-making capabilities of individual agents or nodes, limiting the system's ability to optimize resource allocation based on localized information.

In contrast, DRL offers the potential to overcome these limitations and revolutionize resource allocation in distributed systems. By leveraging DRL, multiple autonomous agents can learn and adapt collaboratively to make decentralized decisions while considering the global system objectives. DRL enables agents to learn from interactions with the environment, continuously update their policies based on feedback, and coordinate their actions to achieve efficient resource allocations. The inherent adaptability of DRL makes it well-suited for dynamic environments, allowing agents to adjust resource allocations in response to changing conditions and evolving demands. DRL can also support scalability in large-scale systems by enabling decentralized decision-making and coordination, thus avoiding the scalability limitations associated with centralized control. Furthermore, DRL has the potential to optimize resource allocation based on diverse and complex objectives, taking into account factors such as fairness, user preferences, and system-wide performance.

Traditional resource allocation methods are often constrained by their inability to adapt to dynamic environments, centralized decision-making, and scalability challenges. offers a promising alternative, with its ability to enable decentralized, adaptive, and scalable resource allocation in modern distributed systems. By leveraging the power of multiple autonomous agents learning and collaborating, DRL has the potential to address the limitations of traditional methods and unlock new possibilities for efficient and intelligent resource allocation.

DRL has emerged as a powerful technique for resource allocation, offering several advantages in decision-making and adaptability. DRL combines DNN with RL algorithms to optimize resource allocation processes and enhance overall efficiency.

First and foremost, DRL excels in making efficient decisions by learning directly from raw data. Unlike traditional approaches that rely on manual rules or heuristics, DRL models adapt and make decisions based on dynamic conditions. The use of DNN enables the processing of vast amounts of information quickly, allowing for the identification of patterns and informed choices. Secondly, Resource allocation problems often occur in uncertain and rapidly changing environments. DRL's real-time learning and adaptation capabilities make it well-suited for such scenarios. Through trial and error, DRL models continually update their strategies, ensuring effective responses to changing conditions and maintaining optimal resource allocation. Thirdly, DRL offers scalability and generalization in addressing complex resource allocation problems. DNN can handle high-dimensional input data, accurately modeling resource allocation scenarios. Additionally, DRL models can generalize their learned policies to unseen situations, allowing for reliable decision-making in unfamiliar contexts. Furthermore, one significant advantage of DRL is its ability to learn from experience. By employing RL techniques, DRL models interact with the environment, receive feedback, and update their policies accordingly. This iterative learning process leads to continuous improvement and optimized resource allocation strategies over time. Lastly, DRL reduces the need for extensive human intervention in resource allocation tasks. Unlike traditional approaches that require manual intervention and domain expertise, DRL autonomously learns optimal resource allocation policies.

This automation saves time and minimizes potential biases and errors associated with human decision-making.

DRL offers several advantages for resource allocation, including efficient decision-making, adaptability to dynamic environments, scalability, learning from experience, and reduced human intervention. By leveraging the power of DNN and RL algorithms, DRL has the potential to revolutionize resource allocation processes across various domains.

In the quest to enhance communication networks' efficiency and user experience, the dynamic allocation of resources like bandwidth, power, and channels emerges as a pivotal challenge. Traditional methods, often static and unable to adapt to fluctuating network conditions, fall short in optimizing network performance. Herein lies the potential of RL, a branch of AI that empowers an agent to learn optimal behaviors through trial and error, guided by feedback from its environment. This section delves into transforming resource allocation challenges into a RL framework, detailing the design of state representations, action spaces, and reward functions to foster effective decision-making.

State Representation: The Foundation

The essence of applying RL in communication networks begins with a precise definition of the state, which reflects the network's current conditions. An effective state representation might encompass:

Network Load: Quantified by metrics like the number of active connections or volume of data traffic, offering a glimpse into the current demand on the network.

Resource Status: Detailed insights into available resources, including bandwidth availability, power levels, and the number of free channels, are crucial.

Performance Indicators: Key QoS metrics such as latency, packet loss rate, and throughput serve as indicators of the network's health and performance.

This multi-dimensional snapshot allows the RL agent to assess the network's immediate needs and resource status, forming the basis for informed decision-making.

Reward Function: Guiding Learning

At the heart of the RL paradigm is the reward function, which steers the agent towards desirable outcomes through positive reinforcement. In the context of network resource allocation, the reward function could be designed to:

Encourage Throughput Maximization: Assigning higher rewards for actions that lead to increased data transmission rates.

Promote Low Latency: Favoring actions that result in reduced transmission delays, thereby improving the user experience.

Foster Fairness: Introducing rewards that motivate the equitable distribution of resources among users, ensuring no user is disproportionately disadvantaged.

Incorporating penalties for negative outcomes, such as dropped connections or QoS violations, further refines the agent's learning process, aligning it with the network's operational goals.

Adopting a RL approach for resource allocation in communication networks offers a dynamic and adaptive solution to optimize network performance. By meticulously defining the states, actions, and rewards, we lay the groundwork for an RL agent capable of navigating the complexities of modern networks, ensuring efficient resource utilization, and enhancing user satisfaction. As this technology matures, it holds the promise of revolutionizing how networks manage their most critical assets in the face of ever-increasing demands.

DRL stands at the forefront of revolutionizing resource allocation in communication networks. Its ability to adapt to changing environments and learn from complex, high-dimensional data makes it an invaluable tool for optimizing network performance and meeting the evolving demands of telecommunications. As research progresses and technology matures, DRL is expected to become a central component of intelligent network management systems, driving innovations in efficiency, reliability, and user experience in communication networks.

3.2.1 DRL-Based User Association

In the field of communication systems, user association refers to the process of determining which users are connected to which base stations (BS) or access points (AP) in a wireless network. This assignment is crucial as it directly impacts the overall performance and efficiency of the communication system. User association plays a vital role in resource allocation, QoS, load balancing, and interference management.

The importance of user association lies in several key aspects. Firstly, user association enables efficient utilization of network resources such as bandwidth, power, and time slots. By assigning users to appropriate BS, the available resources can be optimally distributed, maximizing system capacity and improving network performance. Secondly, user association ensures that users are connected to the BS that can provide the desired QoS. Factors such as throughput, latency, packet loss, and signal strength are taken into consideration when assigning users to BS. Effective user association leads to improved QoS for individual users, resulting in enhanced user experience. Then, user association helps in load balancing by distributing the traffic load across different BS. This prevents congestion and network bottlenecks, ensuring fair resource sharing among users. Load balancing improves overall network stability and reliability. Lastly, user association plays a crucial role in interference management. By appropriately assigning users to BS, interference between users can be minimized. This results in enhanced signal quality and higher data rates.

Traditional user association algorithms are crucial components in wireless communication systems, particularly in cellular networks, where they determine the allocation of users to BS for optimal resource utilization and user experience. These algorithms aim to match users with BSs based on various criteria such as signal strength, load balancing, and QoS requirements. One commonly used traditional user association algorithm is the nearest neighbor (NN) algorithm. In NN, each user is associated with the BS that provides the strongest received signal strength (RSS) at the user's location. This approach is simple and easy to implement but may lead to overloaded BSs and uneven resource utilization, especially in dense networks or areas with non-uniform user distributions. Another traditional algorithm is the proportional fair (PF) algorithm, which seeks to balance fairness and efficiency by considering both the instantaneous channel conditions and the historical data rates of users. PF aims to allocate users to BSs in a way that maximizes the overall system throughput while ensuring a fair distribution of resources among users. However, PF requires more computational complexity compared to NN. Additionally, there are other traditional user association algorithms such as maximum received signal-to-interference-plus-noise ratio (SINR), minimum mean square error (MMSE), and load-aware association. These algorithms consider factors such as interference, channel conditions, and BS load to make more informed decisions regarding user association.

However, traditional user association algorithms in wireless communication systems have certain limitations that can hinder their performance and efficiency. These limitations arise from the complexity and dynamic nature of the wireless environment, as well as the assumptions made in the algorithm design. This sets the stage for the potential of DRL to address these limitations and improve user association strategies. One limitation of traditional user association algorithms is their reliance on predefined metrics or heuristics for decision-making. These metrics may not adequately capture the real-time conditions of the network, leading to suboptimal user assignments. Additionally, traditional algorithms often assume static channel conditions and user demands, which may not reflect the dynamic nature of wireless networks. Another limitation is the difficulty in considering the heterogeneity of BS and their varying capabilities. Traditional algorithms may struggle to effectively allocate users to different types of BS, such as macrocells, microcells, and femtocells, resulting in inefficient resource allocation and suboptimal QoS. Furthermore, traditional user association algorithms typically focus on individual user performance without considering the overall network objectives. This may lead to imbalanced load distribution, causing congestion in some areas and underutilization in others. The lack of coordination among BS can result in increased interference and reduced network capacity.

DRL offers a promising solution to these limitations. By leveraging DNNs and RL principles, DRL algorithms can learn directly from interactions with the wireless environment. This allows them to adapt to changing network conditions and make intelligent decisions based on real-time information. DRL models can capture the complex relationships between user associations, resource allocation, and system-level performance. They can learn to optimize user assignments based on dynamic channel conditions, user demands, and network objectives, leading to improved overall network efficiency, fairness, and QoS. Additionally, DRL has the potential to handle the heterogeneity of BS more effectively. By learning from data, DRL algorithms can adapt to different types of BS and make informed decisions regarding user association, considering the varying capabilities and characteristics of each base station.

One advantage of using DRL for user association is its ability to handle complex and dynamic network conditions. The DNN can capture high-dimensional and non-linear relationships between the state and the optimal action, making it suitable for handling large-scale and heterogeneous wireless networks. Moreover, the network can adapt to changing network conditions without requiring manual adjustments.

DRL holds significant potential for addressing wireless user association problems, revolutionizing how wireless networks allocate users to BS. By leveraging its ability to learn from interactions and optimize decision-making, DRL can offer several advantages in this domain. Applying DRL in wireless user association holds great promise.

Intelligent User Association

DRL can enable intelligent user association decisions by learning from historical network states and user behaviors. Traditional user association approaches often rely on predetermined rules or heuristics, which may not adapt well to dynamic network conditions. DRL, on the other hand, can capture complex patterns and optimize user association in real-time based on current network conditions.

Adaptability to Heterogeneous Environments

Wireless networks are becoming increasingly heterogeneous, with various types of BS and user devices. DRL algorithms have the potential to adapt to these heterogeneous environments and make optimal user association decisions that consider factors like signal quality, traffic load, and mobility patterns. This adaptability can result in improved network performance and better user experiences.

Learning-Based Optimization

DRL can optimize user association decisions through continuous learning and adaptation. As network conditions change over time, DRL models can update their policies based on new observations and interactions. This learning-based optimization can lead to efficient resource allocation, reduced interference, and enhanced overall network capacity.

Handling Complex Interactions

The interactions between users, BS, and the wireless environment are highly dynamic and complex. DRL algorithms can effectively handle these complex interactions by considering the sequential nature of user association decisions. By taking into account the temporal dependencies and long-term rewards, DRL models can optimize user association strategies and mitigate interference issues.

Performance Improvement

Applying DRL in wireless user association has the potential to significantly improve network performance metrics such as throughput, latency, and fairness. By leveraging deep learning techniques, DRL models can capture intricate patterns and make accurate predictions about user demands and network conditions, leading to more efficient user association decisions.

Self-Optimization

DRL has the capability to enable self-optimization of wireless networks. By allowing BS to learn and adapt their user association policies autonomously, DRL can reduce the reliance on centralized management and enable more flexible and efficient network operations. This self-optimization can lead to improved scalability, robustness, and energy efficiency.

In conclusion, traditional user association algorithms have limitations in capturing real-time network conditions, handling heterogeneity, and optimizing overall network performance. DRL offers a promising approach to overcome these limitations by enabling adaptive decision-making based on real-time information and learning from interactions with the wireless environment. The potential of DRL lies in its ability to improve user association strategies, enhance resource allocation, and optimize system-level performance in wireless communication systems.

The basic idea behind using DRL for user association is to frame it as a Markov Decision Process (MDP). In this MDP, the state represents the current network condition and user demand, the action corresponds to the user association decision, and the reward reflects the system performance, such as user satisfaction or resource utilization. To apply DRL, we first need to design a DNN that takes the state as input and outputs the Q-value of each action. The Q-value represents the expected cumulative reward of taking the action under the current state. The network can be trained using Q-learning or policy gradient methods to learn the optimal user association policy. During training, the system interacts with the environment by selecting actions based on the current state and updates the Q-values using the obtained reward. With sufficient training, the network can learn a good user association policy that maximizes the long-term reward. The trained network can then be used to make real-time user association decisions in the communication system.

The state in a communication system embodies the informational context regarding its surroundings. It encompasses critical factors influencing decisions related to user associations, such as channel conditions, SNR, user locations, resource availability, traffic load, and historical data. A comprehensive state allows the system to make informed choices about user associations. For instance, a state representation might comprise a matrix detailing SNR values between users and available BS, offering valuable insights into connection quality crucial for user association decisions. Actions denote the choices available to the communication system concerning user associations. These decisions dictate how users are linked to BS or allocated resources. Actions could involve associating a user with a specific base station, selecting transmission modes, assigning frequency bands, or adjusting power levels. For example, decisions may revolve around identifying the base station providing optimal connectivity for a user based on SNR values or selecting a combination of BS and resources maximizing network capacity while fulfilling individual user requirements. The reward function serves as a guide for the learning process, aiding the optimization of user association decisions. It mirrors the system's objectives and may encompass both system-level and user-centric metrics. At the system level, the function might aim to maximize capacity, minimize interference, or optimize resource utilization. Positive rewards could be allocated for actions leading to higher data rates, reduced latency, or diminished interference. On the other hand, user-centric metrics such as signal quality, fairness, or user experience are also considered. Actions resulting in enhanced signal quality, equitable resource distribution, or improved user experience garner positive rewards. Striking a balance between system-level objectives and user-centric metrics is paramount in designing the reward function. For instance, it could combine system efficiency and user satisfaction metrics, rewarding high data rates, low latency, and fair resource allocation while penalizing actions causing interference or degraded signal quality. By delineating a meticulously crafted state representation, designing suitable actions, and formulating a reward function aligned with system goals and user satisfaction, the user association conundrum can be reframed as a RL challenge. Leveraging RL algorithms enables the communication system to learn optimal user association policies, adapt to evolving conditions, and enhance network performance and user experience iteratively (Fig. 3.1).

In conclusion, the integration of RL into wireless user association presents a compelling opportunity to revolutionize the efficiency and performance of communication networks. With its ability to adapt to changing conditions, optimize resource allocation, and personalize user experiences, RL stands poised to unlock new levels of network intelligence and sophistication. As we continue to advance in this field, we can anticipate a future where wireless communication systems are not only more efficient and reliable but also more responsive to the diverse needs and preferences of users. With ongoing research and development, the full potential of RL in wireless user association is yet to be realized, promising exciting prospects for the evolution of communication technologies in the years to come.

3.2.2 DRL-Based Channel Assignment and Power Allocation

Channel assignment and power allocation are two important techniques used in wireless communication systems to optimize the allocation of radio resources, such as

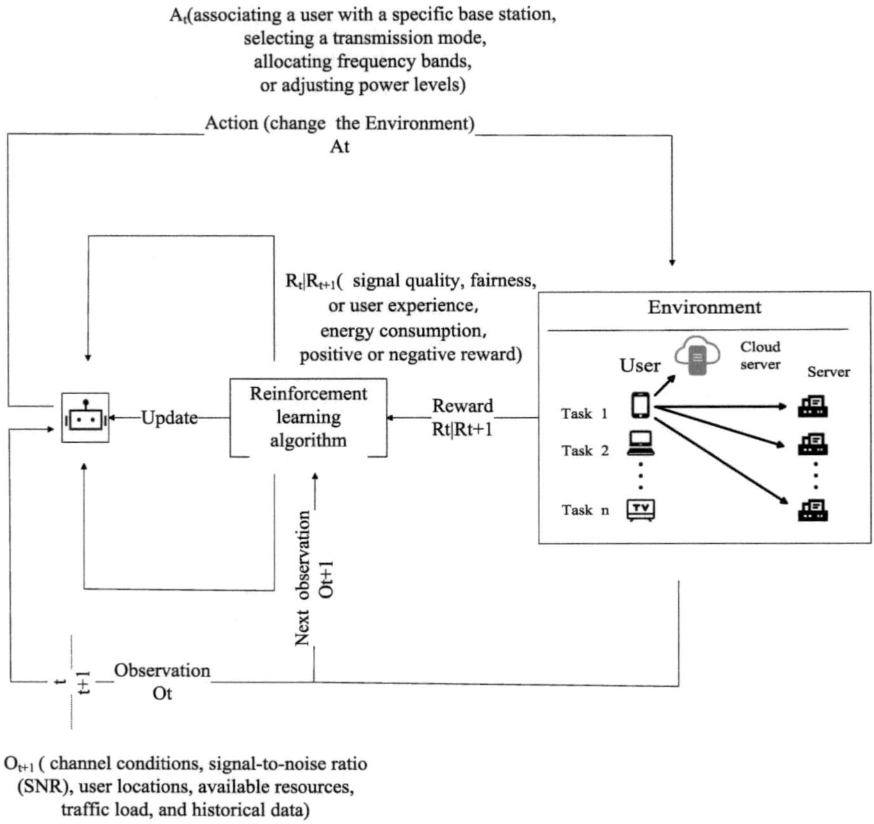

Fig. 3.1 The application of DRL in resource allocation

frequency channels and transmit power, among users to improve system performance. Channel assignment refers to the process of assigning different frequency channels to users in a wireless communication system. In simple terms, it is like dividing a highway into multiple lanes and assigning each lane to a specific user. The main goal of channel assignment is to minimize interference between users and maximize the overall capacity of the system. By allocating different frequency channels to different users, we can ensure that they can communicate without causing significant interference to each other. This helps improve the quality and reliability of wireless connections. Power allocation involves assigning an appropriate amount of transmit power to each user in a wireless communication system. It is like adjusting the volume of your voice when talking to someone—you need to speak louder if the person is far away and softer if they are close. The SNR is a crucial performance metric in wireless communication systems. The SNR can be calculated using the following formula: SNR = (Signal Power) / (Noise Power). Adjusting power allocation to maximize this SNR is a common goal in power allocation strategies to balance the transmission power among users to achieve efficient and reliable communication. In wireless radio propagation models, a path loss model can be used to estimate the distance and channel attenuation between users. Power control based on path loss can be achieved using a formula like:

$$P = P_0 \times d(-\alpha) \tag{3.4}$$

where P is the transmission power, P_0 is the reference power at a certain distance, d is the distance between users, and α is the path loss exponent.

By allocating more power to users with weaker signals or farther distances, we can enhance their signal strength and improve their connectivity. At the same time, users closer to the base station or with stronger signals may require less power, reducing interference and conserving energy.

In the field of communication systems, efficient management of spectrum resources poses several challenges. Spectrum, or the range of frequencies used for wireless communication, is a limited and valuable resource. Here, we will discuss some of the key challenges in spectrum resource management, including spectrum scarcity, increasing user demands, and mutual interference.

Spectrum Scarcity

The availability of usable spectrum is limited, leading to spectrum scarcity. This scarcity arises due to the finite nature of the RF spectrum and the increasing demand for wireless communication services. As more applications and devices require access to the spectrum, the challenge lies in allocating and utilizing the available spectrum efficiently.

Increasing User Demands

With the proliferation of wireless devices, there has been a significant increase in the number of users requiring access to the limited spectrum. Mobile phones, tablets, laptops, IoT devices, and other wireless technologies all compete for spectrum resources. The challenge is to meet the growing demands of users while ensuring fair and equitable access.

Interference Management

In a shared spectrum environment, interference between different users and systems can degrade the quality of communication. Interference occurs when signals from one user or system interfere with the signals of others. As more devices operate in close proximity, the potential for interference increases. Effective interference management techniques, such as frequency planning, power control, and advanced signal processing, are crucial to mitigate this challenge.

Spectrum Fragmentation

Spectrum allocation is typically done in blocks or bands to ensure efficient utilization. However, these allocated bands are not always contiguous, leading to spectrum fragmentation. Fragmentation occurs when spectrum resources are divided into smaller non-contiguous portions, making it difficult for large bandwidth-intensive applications. Managing fragmented spectrum requires intelligent allocation strategies and innovative technologies.

Spectrum Sharing

Traditionally, spectrum has been allocated exclusively to specific services or license holders. However, with the growing demand for spectrum, there is a need for more efficient spectrum sharing mechanisms. DSA and CR technologies are being explored to enable opportunistic spectrum access, allowing multiple users to share the same frequency bands while minimizing interference.

To address these challenges, various strategies are being employed in spectrum resource management. These include spectrum auctions to allocate frequencies, spectrum sharing policies, dynamic spectrum allocation techniques, and the development of advanced technologies like software-defined radio (SDR) and CR.

While various spectrum resource management methods have been developed to tackle the challenges of spectrum scarcity, increasing user demands, and mutual interference, they are not without limitations. Traditional methods like frequency planning, power control, and dynamic spectrum allocation can only optimize spectrum utilization to a certain extent, but may not be able to fully address the complex and dynamic nature of the wireless communication environment. For instance, frequency planning is based on static allocation of spectrum resources that may not adapt well to dynamically changing traffic patterns and network conditions. Power control techniques can mitigate interference but may be limited by the need for coordination among users. Dynamic spectrum allocation can improve spectrum utilization, but may face challenges in implementation due to regulatory and technical complexities.

Therefore, there is growing interest in exploring the application of machine learning techniques, such as DRL, in spectrum resource management. DRL enables autonomous decision-making in a dynamic and uncertain environment, making it a promising approach for addressing the limitations of traditional spectrum management methods. In DRL, an agent learns from experience through trial and error while interacting with its environment. The agent takes actions based on the current state of the environment and receives rewards or penalties based on its decisions. Over time, the agent improves its decision-making ability by maximizing the cumulative reward. In the context of spectrum resource management, DRL can be used to optimize spectrum allocation and access strategies based on real-time feedback from the wireless communication environment. It can also enable autonomous coordination between different users and systems, leading to more efficient spectrum utilization and reduced interference. Several studies have demonstrated the potential of DRL in spectrum resource management. For example, a recent study showed that a DRL-based approach can improve spectrum utilization in a dynamic and heterogeneous wireless network compared to traditional approaches.

While traditional spectrum resource management methods have their limitations in addressing the challenges of spectrum scarcity, increasing user demands, and mutual interference, DRL holds promise in optimizing spectrum utilization and enabling autonomous decision-making in a dynamic wireless communication environment. The future of optimizing spectrum resource management and power allocation in communication systems using DRL holds great promise. By leveraging the power of DRL, we can achieve more efficient allocation of spectrum resources and enhance the overall performance of communication systems. One area of focus is enhanced resource allocation. With DRL techniques, we can dynamically adapt and optimize resource allocation strategies in real-time. This means considering factors such as frequency bands, channel conditions, interference levels, and user demands to make intelligent decisions on how to allocate resources effectively. Intelligent power control is another crucial aspect. By developing DRL algorithms, we can optimize power allocation based on individual user requirements, channel conditions, and network congestion. This leads to improved signal quality, reduced interference, and enhanced energy efficiency in communication systems. Future research can also focus on adaptive learning and decision-making. DRL algorithms can continuously learn and update their policies based on changing network conditions and user behaviors. This allows communication systems to dynamically adjust resource allocation and power control strategies to optimize performance and meet user demands. Moreover, DRL can handle multi-objective optimization problems. By considering trade-offs between different performance metrics like throughput, latency, fairness, and energy efficiency, communication systems can achieve a balance between competing objectives. This enables tailoring resource management and power allocation strategies to specific application requirements.

As communication systems become more complex and heterogeneous, scalability and complexity management are critical. Future research can explore scalable DRL algorithms that efficiently handle large-scale deployments. Techniques such as distributed DRL and hierarchical RL can address the challenges associated with managing complex communication systems. The application of DRL in optimizing spectrum resource management and power allocation in communication systems offers significant potential for enhanced performance and efficiency. Through enhanced resource allocation, intelligent power control, adaptive learning and decision-making, multi-objective optimization, and scalability and complexity management, DRL can revolutionize how we manage and allocate resources in communication systems.

Optimizing spectrum resource management and power allocation in communication systems is crucial to achieving efficient and effective use of available resources. DRL has emerged as a promising approach for addressing this challenge, offering the potential to learn intelligent decision-making policies that can adapt to changing conditions and optimize system performance. In the following paragraphs, we will explore the key concepts and techniques of using DRL for spectrum resource management and power allocation. Firstly, we need to define the problem and formulate it as a RL problem. The primary objective of spectrum resource management is to maximize system throughput while minimizing interference and ensuring fairness among users. The relevant variables that affect spectrum resource management include channel conditions, traffic load, interference levels, and power allocation. We can cast this problem as a MDP, where the state of the system represents the current configuration, actions are the possible decisions to make, and rewards evaluate the quality of the chosen actions. Next, we need to design a suitable DRL architecture for the problem. This typically involves constructing a deep Q-network (DQN) that consists of a state representation, an action space, and a Q-value estimation network. The state representation should incorporate relevant information about the system, such as historical channel information, traffic statistics, and interference levels. The action space should specify the available actions for spectrum allocation and power control, such as selecting frequency bands, adjusting transmit power levels, or allocating resources to different users. The Q-network should estimate the expected cumulative rewards for taking specific actions in a given state. The training process involves collecting experiences, estimating Q-values, and updating the DQN model. During experience collection, we interact with the environment by selecting actions based on an exploration-exploitation strategy and observe the resulting rewards and new states, storing these experiences in a replay buffer. Q-value estimation involves sampling batches of experiences from the replay buffer and computing the Q-value targets using the target network. The network update is performed by minimizing the mean squared error between the predicted Q-values and the target Q-values, using an optimization algorithm like stochastic gradient descent. The target network is updated periodically by copying the weights from the Q-network to improve stability during training. Finally, we can apply the trained DQN model for policy application in real-time scenarios. We observe the current system state and feed it into the trained Q-network to obtain the Q-values for each possible action. We select the action with the highest Q-value (exploitation) or choose a random action with a low probability (exploration) based on the chosen exploration-exploitation strategy. We then apply the selected action to the communication system, adjusting spectrum allocation and power levels accordingly. The policy should continuously monitor the system's state, repeat the decision-making process, and adapt as the environment evolves.

In conclusion, leveraging DRL for spectrum resource management and power allocation offers a promising future for communication systems. By further advancing research in these directions, we can unlock the full potential of DRL to optimize resource utilization, enhance system performance, and provide seamless connectivity for a wide range of applications.

3.2.3 DRL-Based Energy Transfer and Harvesting

In a communication system, the transfer and collection of signal energy play a crucial role in ensuring reliable and efficient data transmission. Signal energy transfer refers to the process of transmitting information-carrying signals from a source to a destination, while signal energy harvesting involves collecting and utilizing ambient or transmitted energy to power wireless devices. Both aspects are essential for maintaining continuous and sustainable communication in various applications.

Energy Transfer

- Reliable Data Transmission. Signal energy transfer is vital for ensuring that information-carrying signals can reach their intended destination with sufficient strength and quality. Proper energy transfer mechanisms help mitigate signal degradation and ensure reliable data transmission across different distances.
- Wireless Communication. In wireless communication systems, such as cellular networks, Wi-Fi, and bluetooth, efficient signal energy transfer enables seamless connectivity and data exchange between devices without the need for physical connections. This is essential for modern mobile and IoT applications.
- Remote Sensing and Monitoring. Signal energy transfer is critical for remote sensing and monitoring systems, such as environmental monitoring, smart grids, and industrial automation. It allows for the transmission of sensor data over long distances, enabling real-time monitoring and control of distributed systems.

Energy Harvesting

- Energy Sustainability. Signal energy harvesting provides an alternative means of powering wireless devices by capturing ambient energy from the environment, such as light, vibration, or RF signals. This approach promotes energy sustainability and reduces reliance on traditional power sources, particularly in remote or off-grid locations.
- Autonomous and IoT Devices. Signal energy harvesting enables the deployment of autonomous and IoT devices that can operate without the need for frequent battery replacements or external power sources. This is particularly beneficial for applications such as wireless sensor networks, wearable electronics, and remote monitoring systems.
- Environmental Adaptability. By harnessing ambient energy, signal energy harvesting allows devices to adapt to diverse environmental conditions and operate in locations where traditional power sources may be limited or unavailable. This flexibility enhances the deployment and usability of wireless communication devices in various scenarios.

Signal energy transfer and energy harvesting are indispensable for maintaining reliable communication, promoting energy sustainability, and enabling the deployment of wireless devices in diverse applications, ranging from smart cities to wearable electronics. These technologies are fundamental to the advancement of wireless communication systems and the broader ecosystem of connected devices.

Traditional methods of signal energy transfer and energy harvesting have played a crucial role in communication systems and power supply. However, they are not without limitations that can impact their effectiveness in various scenarios. One significant limitation is signal attenuation. As signals travel through the air or other mediums, their strength diminishes, leading to reduced signal quality and reliability over long distances. This attenuation can limit the effective range of communication systems and hinder reliable data transmission. Another limitation lies in the environmental dependency of many energy harvesting methods. Solar panels, for example, require adequate sunlight, while RF energy harvesting relies on the availability of RF signals. These dependencies restrict the applicability of these methods in environments with limited or fluctuating energy sources. Furthermore, energy losses occur throughout the process of signal transmission and energy conversion. Factors such as resistance, noise, and inefficiencies in power conversion contribute to these losses, reducing the overall effectiveness of traditional energy transfer and harvesting methods. As a result, devices reliant on harvested energy may have limited longevity.

To address these limitations, DRL shows promise. This branch of artificial intelligence leverages advanced algorithms and training models to optimize signal transmission parameters and adapt to environmental changes. DRL can intelligently adjust signal transmission parameters, such as power levels and modulation schemes, to minimize energy loss and attenuation over distance. By learning from environmental data, it can make real-time decisions on when and how to harvest energy, effectively addressing the issue of environmental dependency. By overcoming the limitations of traditional methods, DRL has the potential to improve the efficiency and performance of communication systems and energy-harvesting devices. Its intelligent decision-making and adaptive optimization can enhance signal quality, extend the range of communication systems, and maximize energy efficiency.

Traditional signal energy transfer and energy harvesting methods have limitations such as signal attenuation, environmental dependency, and energy loss. However, DRL offers a promising solution by enabling intelligent decision-making and adaptive optimization. By addressing these limitations, DRL can enhance the efficiency and performance of communication systems and energy-harvesting devices in various scenarios.

Firstly, DRL can effectively mitigate signal attenuation in energy transfer. Traditional methods often suffer from signal loss over long distances, impacting the reliability and range of communication systems. By optimizing transmission parameters such as power levels and modulation schemes, DRL can minimize energy loss over distance, extending the effective range of communication systems and enhancing data transmission reliability. Secondly, DRL exhibits adaptability to environmental changes in energy harvesting. Many traditional energy harvesting methods are dependent on specific environmental conditions, such as sunlight or RF signals.

DRL can learn from environmental data and make real-time decisions on energy harvesting, effectively addressing these environmental dependencies. This adaptability optimizes energy harvesting processes, maximizing energy efficiency and reducing reliance on external power sources. Furthermore, DRL can address energy losses throughout the energy transfer and conversion processes. By optimizing power conversion processes, DRL can reduce energy losses stemming from resistance, noise, and inefficiencies. These improvements enhance the overall efficiency and functionality of devices reliant on harvested energy, leading to longer device lifetimes and reduced energy waste. Lastly, DRL's intelligent decision-making and adaptive optimization capabilities contribute to enhancing the efficiency and performance of communication systems and energy-harvesting devices. Its ability to learn and adapt from environmental data allows for dynamic adjustments to energy transfer and harvesting processes, leading to improved reliability and energy utilization.

DRL has emerged as a powerful tool for optimizing signal energy transfer through wireless networks. By leveraging advanced algorithms and DNN, this approach enables efficient decision-making by the agents involved in transmitting and receiving signals.

To begin, let's consider a scenario where we have multiple transmitters and receivers in a wireless network. The objective is to maximize the amount of signal energy transferred from the transmitters to the receivers while considering various factors like channel conditions, interference, and power constraints. In this context, each transmitter acts as an agent, and the environment comprises the wireless channel and other agents. The agents learn how to make decisions that optimize the energy transfer through a process called DRL.

The first step is to define the state representation. The state provides information about the current environment to the agent, including the channel conditions, interference levels, and the energy levels at the transmitter and receiver. This information helps the agent make informed decisions. The state representation is designed based on the available information and the specific requirements of the energy transfer task. Next, we define the action space. The action represents the decision made by the agent regarding how to transmit the signal. It could include adjusting the transmit power, choosing the frequency or time slots for transmission, or other actions that affect signal transmission. The action space should be carefully designed to cover a range of possible actions while remaining computationally feasible. Once the state and action spaces are defined, we establish the reward function. The reward serves as feedback to the agent and guides its learning process. In the context of signal energy transfer, the reward function should encourage the agent to maximize the energy transferred while accounting for factors like interference and power efficiency. For example, the agent can receive a positive reward for successfully transferring energy and a negative reward proportional to interference caused or the power consumed. By designing an appropriate reward function, we can train the agent to make decisions that align with our objectives. With the RL task defined, we can train the agent using DNN. The agent interacts with the environment, observing the state, taking actions, and receiving rewards. DNN approximate the value or policy function, mapping states to actions or action probabilities. Through techniques like Q-learning or policy gradients (PG), the agent learns to make better decisions over time.

During training, the agent explores different actions and learns from the feedback received through rewards. By iteratively adjusting the neural network parameters, the agent gradually improves its decision-making capabilities and converges to an optimal policy for signal energy transfer. Once trained, the agent can be deployed in real-world scenarios to facilitate efficient signal energy transfer in wireless networks. The agent intelligently adjusts its transmission strategy based on the observed state to maximize energy transfer while minimizing interference and optimizing power usage.

DRL offers a powerful algorithmic framework for optimizing signal energy transfer in wireless networks. By defining the problem as a RL task and leveraging DNN, we can train agents to make informed decisions that maximize energy transfer efficiency. This approach has the potential to significantly improve the performance of wireless networks and enable efficient and reliable signal energy transfer. DRL has the potential to revolutionize the field of signal energy transmission and collection, offering numerous exciting applications and advancements. With its ability to optimize decision-making processes in complex and dynamic environments, DRL can greatly enhance the efficiency and effectiveness of signal energy management.

One significant application lies in optimizing wireless power transfer systems. Wireless power transfer enables devices to receive energy without physical connections, offering convenience and flexibility. However, efficiently managing and optimizing the transfer process is crucial. DRL algorithms can train agents to make intelligent decisions on power transfer parameters, such as distance, direction, and frequency, to maximize energy transfer efficiency while minimizing losses and interference. Another exciting area of application is in wireless sensor networks. These networks often consist of numerous battery-powered sensors that collect and transmit data. Energy consumption is a significant concern in such networks, as batteries may need frequent replacement or recharging. DRL can be used to optimize the energy consumption of sensors by dynamically adjusting their transmission power, scheduling data transmissions, and optimizing routing decisions. This can lead to prolonged battery life, reduced maintenance costs, and improved network performance. Furthermore, DRL can contribute to the development of energy-efficient communication protocols. Wireless communication protocols, such as Wi-Fi or bluetooth, play a crucial role in transmitting and receiving signals. Optimizing these protocols for energy efficiency is essential to conserve power and extend device battery life. DRL agents can learn to adapt transmission parameters, such as modulation schemes and transmission power, based on the current channel conditions and energy constraints. This can result in significant energy savings without compromising communication performance. Additionally, DRL can aid in optimizing energy harvesting systems. Energy harvesting technologies, like solar panels or kinetic energy harvesters, allow devices to generate power from their surroundings. However, the availability of harvested energy varies, making efficient energy management challenging. By training agents using DRL, devices can intelligently decide when to harvest energy, how much to store, and when to utilize the stored energy. This can maximize the utilization of available energy sources and ensure continuous operation. Moreover, DRL can play a crucial role in managing interference in signal energy transmission. Interference is a common challenge in wireless communication, affecting signal quality and energy efficiency. DRL agents can learn to mitigate interference by dynamically adjusting transmission parameters, optimizing frequency allocations, or employing interference cancellation techniques. This can lead to improved signal quality, enhanced energy transfer efficiency, and better overall network performance.

In conclusion, the application prospects of DRL in the field of signal energy transmission and collection are vast and promising. From optimizing wireless power transfer systems and energy consumption in sensor networks to enhancing communication protocols and managing interference, DRL offers exciting opportunities for more efficient and sustainable utilization of signal energy. As research and development progress in this field, we can expect to witness significant advancements that will revolutionize how we transmit, collect, and manage signal energy in various applications and industries.

3.2.4 MADRL-Based Resource Management

Resource management plays a crucial role in distributed systems and networks, encompassing a wide range of activities aimed at optimizing the allocation and utilization of resources to ensure efficient and reliable system operation. The importance of effective resource management in these environments cannot be overstated, as it directly impacts system performance, scalability, and user experience. One key aspect of resource management in distributed systems and networks is the allocation of computing resources, such as CPU, memory, and storage. Efficiently allocating these resources is essential for ensuring that tasks and applications can run smoothly without contention or performance degradation. Proper resource allocation directly contributes to system responsiveness, throughput, and overall user satisfaction. Furthermore, in the context of network resource management, efficient allocation of bandwidth and network capacity is critical for ensuring optimal data transmission and communication. By managing network resources effectively, it is possible to minimize congestion, reduce latency, and ensure equitable access to resources for all users. This is particularly important in today's interconnected world, where seamless and reliable network connectivity is essential for various applications and services. Another crucial aspect of resource management in distributed systems is the management of distributed data. This includes tasks such as data storage, replication, and consistency management. Effective management of data resources ensures data availability, reliability, and integrity, which are fundamental requirements for many distributed applications and services.

In addition to resource allocation and data management, proper load balancing is also an important component of resource management in distributed systems. Load balancing aims to evenly distribute processing and communication tasks across available resources to prevent bottlenecks and overloads. Effective load balancing strategies contribute to improved system performance, fault tolerance, and scalability. Moreover, security and access control mechanisms are integral parts of resource management in distributed systems and networks. Properly managing access to resources and implementing security measures are essential for protecting sensitive data, preventing unauthorized access, and ensuring the integrity and confidentiality of system resources.

Overall, effective resource management in distributed systems and networks is essential for optimizing system performance, ensuring reliable operation, and providing a satisfactory user experience. Through efficient resource allocation, network capacity management, data management, load balancing, and security measures, organizations and service providers can maximize the utilization of resources, improve system responsiveness, and maintain the integrity and security of distributed systems and networks.

Traditional resource management methods have been the backbone of system administration for many years, but they have some limitations. These methods are typically rule-based and rely on predefined policies and procedures to allocate resources. While these methods can be effective in simple systems with predictable workloads, they struggle to adapt to complex and dynamic environments. One of the key limitations of traditional resource management is that it struggles to handle non-deterministic workloads. Workloads that are unpredictable or change frequently can cause contention or underutilization of resources. This can lead to poor system performance, increased latency, and even system failure. Moreover, traditional resource management methods do not take into account individual user requirements or preferences. Resource allocation decisions are based on pre-defined policies and procedures rather than specific user needs. This can lead to suboptimal resource utilization and reduced user satisfaction.

To address these limitations, recent research has focused on using DRL techniques to optimize resource management. DRL is a type of machine learning that uses trial-and-error to learn optimal resource allocation strategies from experience. Using DRL, it is possible to dynamically adapt resource allocation strategies based on real-time workload conditions and user demands. DRL algorithms can take into account various factors such as system performance metrics, user preferences, and workload patterns to make informed decisions on resource allocation. Moreover, DRL can handle non-deterministic workloads more effectively than traditional methods. By continuously learning and adapting to changing conditions, DRL algorithms can optimize resource allocation strategies to ensure efficient utilization of resources. In addition, DRL can handle multi-objective optimization problems, allowing organizations to balance competing goals such as performance, energy efficiency, and user satisfaction. By considering trade-offs between different objectives, DRL can help organizations achieve a balance between resource utilization and user experience.

DRL holds great promise for optimizing resource management in complex and dynamic environments. By leveraging the power of machine learning techniques, organizations can achieve more efficient resource utilization, improved system performance, and enhanced user satisfaction.

One key advantage of multi-agent DRL is its ability to capture the complex interactions and dependencies among agents. In distributed systems, different agents often share resources and influence each other's performance. By using DRL, agents can learn to make decisions that consider the impact on other agents and the overall system. This collaborative decision-making enables agents to better coordinate their actions, leading to improved resource utilization and system performance. Another advantage of multi-agent DRL is its ability to handle dynamic and non-stationary environments. In distributed systems, resource availability and workload conditions can change rapidly. Traditional approaches struggle to adapt to these changes effectively. However, with DRL, agents can continuously learn and update their policies based on real-time feedback, enabling them to adapt to evolving conditions. This adaptive capability allows agents to optimize resource allocation even in highly dynamic environments. Furthermore, multi-agent DRL promotes scalability and flexibility in distributed resource management. As the number of agents increases, traditional methods often suffer from communication overhead and coordination challenges. In contrast, multi-agent DRL allows agents to learn and act locally, reducing the need for extensive communication and central coordination. This decentralized approach enables more scalable and flexible resource management, particularly in large-scale distributed systems. Additionally, multi-agent DRL facilitates the exploration of diverse resource allocation strategies. Traditional methods often rely on predefined rules and policies, limiting the exploration of alternative approaches. In contrast, DRL enables agents to explore different actions and learn optimal strategies through trial-and-error. By encouraging exploration, multi-agent DRL can uncover novel and more efficient resource allocation policies that may have been overlooked by traditional methods.

Multi-agent DRL offers significant advantages in handling distributed resource management problems. By enabling collaborative decision-making, adapting to dynamic environments, promoting scalability and flexibility, and facilitating exploration, multi-agent DRL can improve resource utilization, system performance, and overall efficiency in distributed systems.

To transform resource management problems into MADRL tasks, we model the problem as a MDP. A MDP is a mathematical framework that represents a decision-making problem as a set of states, actions, and rewards. At each time step, the agents observe the current state of the system (i.e., resource availability, workload demand, etc.), take actions based on their policy (i.e., an algorithm that maps states to actions), and receive a reward that reflects the quality of their action. The goal of the agents is to learn a policy that maximizes their expected cumulative reward over time. The state space in resource management can be designed to include relevant information that agents need to make decisions. For example, the state space could include the current resource allocation, the workload demand, user preferences, system performance metrics, and so on. The action space represents the set of actions that each agent can take at each time step. In resource management, actions could include allocating resources to different users or applications, adjusting system parameters, or requesting additional resources. The reward function specifies the goal of the agents and provides feedback on their actions. In resource management, the reward function could be designed to maximize resource utilization, minimize latency, balance the workload across the system, or optimize energy efficiency. Once the state, action, and reward functions are defined, we can use DRL to optimize resource management. DRL is a combination of RL and deep learning that involves training a neural network to approximate the optimal policy that maximizes expected cumulative reward. The neural network takes as input the current state and outputs a probability distribution over the set of possible actions. The policy is learned by iteratively updating the weights of the neural network based on the observed rewards and resulting state transitions.

One advantage of using DRL is its ability to handle high-dimensional state and action spaces. This allows agents to observe and influence complex system states, leading to more effective resource management. Additionally, DNNs can learn representations that capture the underlying structure of the state space, enabling more efficient decision-making. However, there are also challenges in applying DRL to resource management, such as dealing with non-stationary environments, handling communication overhead in large-scale systems, and ensuring fairness in resource allocation. Transforming resource management problems into MADRL tasks and using DRL to optimize decision-making can lead to more efficient and fair resource allocation. By modeling the problem as an MDP and designing the state, action, and reward functions appropriately, agents can learn to make effective decisions in dynamic and complex environments.

Multi-agent deep reinforcement learning MADRL has emerged as a promising approach for addressing the challenges of distributed resource management. By enabling multiple agents to learn and adapt collaboratively, MADRL has the potential to revolutionize how resources are allocated in complex systems.

Efficient Resource Allocation

One of the key advantages of MADRL in resource management is its ability to optimize resource allocation efficiently. Traditional approaches often suffer from suboptimal decision-making due to limited coordination and information exchange between agents. With MADRL, agents can learn to make decisions based on a global perspective, taking into account the system-wide resource availability, workload demands, and user preferences. By leveraging DNN, MADRL can handle high-dimensional state and action spaces, enabling more effective and accurate resource allocation.

Scalability in Large-Scale Systems

As systems become increasingly large-scale and distributed, resource management becomes more challenging. MADRL offers a scalable solution by allowing agents to operate autonomously and make localized decisions while still considering the global system objectives. Through decentralized decision-making and coordination, MADRL can effectively manage resources across multiple nodes, devices, or even edge computing environments. This scalability makes MADRL particularly well-suited for modern distributed systems with diverse resource requirements.

Adaptability to Dynamic Environments

Distributed resource management often faces dynamic and changing environments. Workload demands fluctuate, resource availability varies, and system conditions evolve over time. MADRL excels in such environments due to its ability to learn and adapt. Agents can continuously update their policies and adjust resource allocations based on real-time observations and rewards. This adaptability allows MADRL to handle unexpected changes, optimize resource utilization, and maintain system performance even in the face of varying conditions.

Fairness and Resource Balancing

Ensuring fairness in resource allocation is a critical aspect of distributed systems. MADRL provides a framework to achieve fairness by considering the preferences and requirements of different users or applications. Agents can learn to balance the workload across the system, prioritize certain tasks based on user-defined criteria, or dynamically allocate resources based on changing priorities. By incorporating fairness considerations into the reward function, MADRL can optimize resource allocation while ensuring equitable access to resources for all participants.

Collaborative Decision-Making

Another exciting aspect of MADRL in resource management is its potential to enable collaborative decision-making. Agents can learn to communicate, negotiate, and cooperate to achieve better resource allocation outcomes. This collaboration allows agents to share information, coordinate actions, and collectively optimize the system performance. By leveraging the power of multiple agents working together, MADRL can unlock new possibilities for efficient and intelligent resource management.

The future of MADRL in distributed resource management holds great promise. Through efficient resource allocation, scalability in large-scale systems, adaptability to dynamic environments, fairness considerations, and collaborative decision-making, MADRL can revolutionize how resources are managed in complex distributed systems. As researchers continue to advance MADRL algorithms and techniques, we can expect significant advancements in optimizing resource allocation and improving system performance in diverse domains such as cloud computing, IoT, and edge computing. With its ability to leverage the collective intelligence of multiple agents, MADRL has the potential to transform the way we manage and utilize resources in the digital age.

3.2.5 DRL-Based Task Offloading

Edge computing refers to a distributed computing architecture that brings computation and data storage closer to the edge of the network, rather than relying on centralized cloud servers. It aims to process and analyze data locally, at or near the source of data generation, instead of sending all the data to a remote data center. Task offloading, also known as task migration or task offloading, is a technique used in edge computing where specific tasks or workloads are shifted from a local device or edge node to a remote server or cloud infrastructure. This is done to optimize resource utilization, improve performance, reduce latency, and enhance energy efficiency.

The concept of edge computing has gained significant importance due to several reasons. Firstly, the exponential growth of IoT devices and their increasing reliance on real-time data processing demands faster response times. By moving computation closer to the edge devices, edge computing minimizes the latency caused by sending data back and forth to the cloud, enabling real-time and near real-time applications. Secondly, edge computing addresses the challenge of limited bandwidth and network congestion. With the rapid increase in data volume, transmitting all the data to a centralized cloud can overload the network infrastructure. By performing data processing and analysis at the edge, only relevant and summarized data needs to be transmitted, reducing network congestion and optimizing bandwidth usage. Furthermore, edge computing enhances data privacy and security. By keeping sensitive data local, it reduces the risk of data breaches and ensures compliance with data privacy regulations. Additionally, edge computing enables offline operation and resilience, allowing devices to continue functioning even when the connection to the cloud is disrupted. Lastly, edge computing offers scalability and cost-effectiveness by distributing computational resources closer to the point of need. Instead of relying solely on powerful and expensive centralized servers, edge devices can contribute to the overall computing capacity, resulting in improved efficiency and reduced costs.

The concept of edge computing, as outlined in the previous paragraph, addresses various challenges related to real-time data processing, bandwidth optimization, data privacy, and scalability. However, despite its advantages, implementing edge computing solutions also introduces its own set of challenges. One such challenge lies in the effective uninstallation process of applications or tasks deployed at the edge. This process encounters various obstacles, including latency, bandwidth limitations, resource management issues, dependency resolution complexities, and error handling difficulties. These challenges are inherent to the dynamic nature of edge computing environments and must be carefully addressed to ensure the smooth removal of tasks. The uninstallation process encounters various challenges that can hinder its effectiveness and efficiency. These challenges include latency, bandwidth limitations, resource management issues, dependency resolution complexities, and error handling difficulties. Addressing these challenges is crucial to ensure a smooth and successful task removal process. First of all, latency presents a significant hurdle during uninstallation, referring to delays in data processing or transfer. High latency can lead to inefficiencies and delays in completing the uninstallation, impacting overall task management. Minimizing latency involves optimizing network infrastructure, utilizing efficient communication protocols, and appropriately prioritizing tasks. Then bandwidth limitations also pose a challenge, restricting data transfer rates during the uninstallation process. Limited bandwidth can slow down uninstallation, especially for tasks with large files or complex configurations, potentially disrupting other network activities. Effective bandwidth management strategies such as traffic shaping and QoS mechanisms are essential to address this challenge. Thirdly, proper resource management is critical for allocating memory, processing power, and disk space efficiently during uninstallation. Inadequate resource allocation can result in performance issues, system instability, or incomplete task removal. Implementing resource monitoring, dynamic allocation, and load balancing techniques can optimize resource utilization and improve the uninstallation process. Moreover, dependency resolution is another challenge, as uninstalling a task may involve resolving dependencies with other software components or libraries. Mishandling dependencies can lead to system instability or incomplete uninstallation. Employing tools for dependency analysis, version compatibility checks, and rollback mechanisms is essential for identifying and resolving dependencies effectively. Lastly, error handling is crucial for managing unexpected issues during the uninstallation process, such as file conflicts, permission problems, or system errors. Robust error handling mechanisms, including detailed logging, defined rollback procedures, and user notifications, are necessary to address errors promptly and minimize disruptions.

Traditional methods often face significant limitations when addressing challenges such as latency, bandwidth limitations, resource management issues, dependency resolution complexities, and error handling difficulties in task offloading scenarios. Latency, referring to the delay in data processing and transfer, can impede the efficiency of traditional methods, particularly in time-sensitive applications where real-time responsiveness is crucial. Bandwidth limitations further compound this issue by restricting the volume of data that can be transmitted within a specific timeframe, leading to potential congestion and performance bottlenecks. Resource management poses another obstacle for traditional task offloading approaches. Allocating and coordinating resources across distributed systems manually can be complex and prone to inefficiencies, resulting in suboptimal resource utilization and overall system performance. Dependency resolution complexities also challenge traditional methods, as managing the interdependencies between different tasks and ensuring their proper execution order can be intricate and error-prone. This can lead to operational disruptions and hinder overall system reliability and performance. Moreover, traditional methods may struggle with error handling, relying on predefined rules and algorithms that may not adapt well to unexpected situations or evolving conditions. This lack of flexibility can result in subpar error recovery mechanisms and potentially impact system stability and reliability.

In contrast, DRL offers promising potential in addressing these challenges in task offloading. By utilizing DNNs and RL algorithms, DRL can autonomously learn and optimize strategies for task offloading in dynamic environments. It can adapt to varying latency and bandwidth conditions, dynamically adjust resource allocation, learn to resolve dependencies efficiently, and enhance error handling mechanisms through continuous learning and improvement. The adaptive nature of DRL enables systems to evolve and improve their decision-making processes over time, offering a robust solution to the limitations of traditional task offloading methods. By learning from interactions and optimizing performance iteratively, DRL has the capacity to revolutionize task offloading and overcome the complexities inherent in modern distributed computing environments.

One key advantage of DRL in task offloading is its adaptability. DRL algorithms excel at adapting to dynamic and uncertain environments, allowing for real-time optimization of task offloading strategies. This adaptability enables the system to adjust to changing network conditions and workload variations efficiently, ensuring optimal performance under varying circumstances. Moreover, DRL enables complex decision-making processes in task offloading scenarios. By considering multiple factors such as computational load, network latency, and energy consumption, DRL can intelligently allocate tasks to achieve optimal outcomes. The ability to make sophisticated decisions based on learned policies enhances the efficiency and effectiveness of task offloading systems. Optimal resource allocation is another significant benefit of applying DRL to task offloading. DRL models can continuously learn from experience and feedback, refining the offloading strategy over time. This iterative learning process leads to more efficient resource utilization, reduced latency, and overall improved system performance in task offloading operations. Furthermore, the scalability and generalization capabilities of DRL make it well-suited for diverse task offloading applications. DRL models can scale to handle various tasks and environments, while also generalizing learned policies to new, unseen situations. This flexibility enhances the adaptability of task offloading systems, enabling them to perform effectively across a wide range of scenarios. Finally, the real-time learning aspect of DRL is crucial for dynamic task offloading scenarios. DRL allows for continuous learning and decision-making, enabling systems to quickly adapt to changing conditions without manual intervention. This real-time adaptability ensures responsiveness and efficiency in task offloading operations, particularly in environments where conditions may change rapidly.

In conclusion, by leveraging the adaptability, complex decision-making capabilities, optimal resource allocation, scalability, generalization, and real-time learning provided by DRL, task offloading systems can achieve enhanced performance, efficiency, and responsiveness across diverse applications and environments (Fig. 3.2).

DRL has emerged as a powerful technique to optimize various complex tasks. Central to the task offloading problem in edge computing is the need to make real-time decisions regarding task allocation, destination edge nodes, and resource utilization. This necessitates a comprehensive formulation of the problem within the framework of RL, entailing the definition of the state space, action space, and reward function to guide the learning process. Effective representation of the environment state is paramount for enabling the RL agent to make informed decisions. This involves capturing pertinent information such as workload status at edge nodes, network congestion levels, task requirements, and user QoS preferences. Through judicious employment of feature engineering techniques, raw data is distilled into a compact

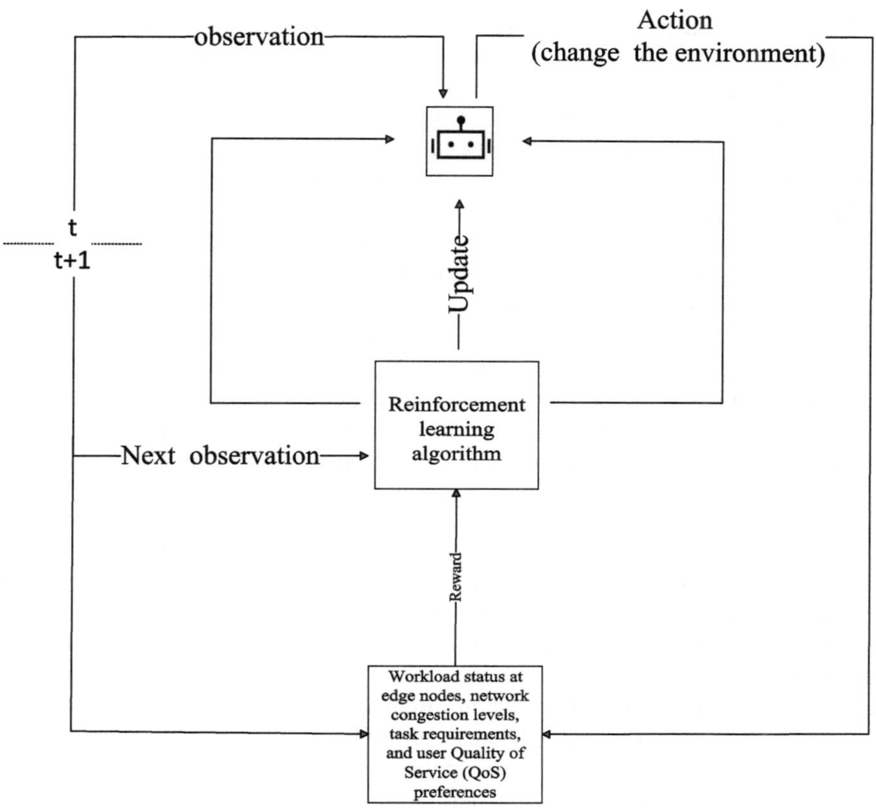

Fig. 3.2 The application of deep reinforcement learning in task offloading

and informative state representation. The action space, encompassing the permissible decisions the RL agent can undertake at each time step, must be meticulously crafted to ensure feasibility and scalability while accommodating the dynamic nature of edge environments. Likewise, the design of an appropriate reward function is crucial in steering the RL agent towards desired optimization goals, whether minimizing latency, maximizing throughput, or optimizing energy consumption. The reward function plays a crucial role in shaping the behavior of the agent. It quantifies the desirability of different actions taken by the agent and guides it towards making optimal decisions. In the context of task offloading, a suitable reward function could consider multiple objectives, such as minimizing response time, maximizing energy efficiency, and maintaining acceptable QoS. The reward function can be designed as a weighted sum of these objectives, with the weights reflecting their relative importance. For example, the reward function can be defined as:

Reward $= \alpha \times$ Response Time $- \beta \times$ Energy Consumption $+ \gamma \times$ QoS

Here, α, β, and γ are the weights assigned to each objective. Adjusting these weights allows us to prioritize different goals based on the specific requirements of the task offloading scenario.

Selecting an apt RL algorithm is pivotal in effectively solving the task offloading problem. Common choices include Q-learning, DQN, and Policy Gradient methods like Proximal Policy Optimization (PPO) or Trust Region Policy Optimization (TRPO). The chosen algorithm should adeptly navigate the high-dimensional state and action spaces inherent in edge computing environments while effectively balancing exploration and exploitation. During the training phase, the RL agent interacts with the environment, assimilating states, executing actions, receiving rewards, and updating its policy iteratively based on the accrued feedback. Training data may be sourced from simulations or historical data gleaned from real-world edge deployments. Over time, the agent hones its ability to optimize task offloading decisions, maximizing cumulative rewards. Upon completion of training, the RL agent is deployed in a real-world edge computing environment or a realistic simulation to evaluate its efficacy. Monitoring mechanisms are established to scrutinize the agent's decision-making process, track system performance, and detect any deviations from expected behavior. Continuous evaluation and refinement are indispensable to ensure the agent's adaptability in dynamic edge environments. Integration with existing edge infrastructure is imperative for the practical deployment of the RL-based task offloading framework. This may entail developing APIs or interfaces to facilitate seamless interaction between the RL agent and edge nodes, ensuring compatibility with prevailing protocols and systems. Regular monitoring and maintenance are indispensable to uphold the effectiveness of the RL-based task offloading system. This encompasses updating the agent's policy to align with evolving requirements and conditions, addressing edge node failures or resource constraints, and adapting to shifts in workload patterns or network dynamics.

In conclusion, the potential applications of DRL for task offloading are vast and diverse, with implications for mobile edge computing, industrial automation, smart cities, and beyond. As research and development in this field continue to advance, we can expect to see innovative solutions that optimize resource utilization, enhance system performance, and enable intelligent decision-making in dynamic and complex environments. The future is bright for the integration of DRL into task offloading, promising significant advancements across various industries and domains.

3.2.6 DRL-Based Load Balancing

In the context of computer networking, load balancing refers to the distribution of incoming network traffic or workload across multiple servers or resources. The primary purpose of load balancing is to ensure that no single server or resource becomes overwhelmed with traffic, thereby optimizing resource utilization, maximizing throughput, and minimizing response time. When a client initiates a request, such as accessing a website or an application, the load balancer intercepts the request and forwards it to one of several backend servers based on a predetermined set of algorithms and criteria. These algorithms may take into account factors such as server availability, current workload, and overall system performance. By distributing the incoming requests across multiple servers, load balancing helps to prevent any one server from becoming a bottleneck, thus improving the overall reliability and scalability of the system. Furthermore, load balancing can also enhance fault tolerance and resilience. If one server fails or becomes unresponsive, the load balancer can redirect traffic to other healthy servers, ensuring continuous availability of services. Additionally, load balancing facilitates horizontal scalability, allowing additional servers to be easily added to the system to handle increased traffic without requiring significant changes to the overall architecture.

Traditional load balancing methods are fundamental techniques employed in computer networking to effectively distribute incoming network traffic among multiple servers. These methods play a critical role in optimizing resource utilization and ensuring high availability of services. Several common traditional load balancing methods include round robin, Least Connections, Weighted Round Robin, IP hash, and URL hash. The round robin method involves sequentially assigning incoming requests to each server in a rotating fashion. This approach ensures an even distribution of the workload across all servers, providing a fair allocation of incoming traffic. Meanwhile, the Least Connections algorithm directs incoming requests to the server with the fewest active connections at any given time. By doing so, it helps balance the load by sending new requests to servers that are less busy. In the Weighted Round Robin method, servers are assigned weights based on their capacity or performance. This means that servers with higher weights receive a proportionally greater share of incoming requests, taking into account their individual capabilities. Additionally, the IP hash method utilizes the source IP address of incoming requests to determine which server will handle the request. Requests from the same IP address are consistently directed to the same server for improved continuity. Similarly, the URL Hash method leverages the requested URL to decide which server should process the request. Requests for specific URLs are routed to designated servers based on the hash value of the URL. Despite their widespread use, traditional load balancing methods have inherent limitations. For example, they may not dynamically consider the real-time loads on servers, potentially leading to uneven distribution of workload and suboptimal resource utilization. Furthermore, these methods may struggle to adapt to sudden traffic spikes or failover scenarios, which can impact the overall performance and reliability of the system.

However, traditional load balancing methods, such as round-robin, least connections, and IP hash algorithms, have been widely used to distribute incoming network traffic among servers in a balanced manner. These methods operate based on predetermined rules or criteria to determine how requests should be directed to different servers. While effective in many scenarios, traditional load balancing approaches have inherent limitations that can impact their performance and adaptability in dynamic network environments. One significant limitation of rule-based load balancing algorithms is their static nature, meaning they do not easily adjust to changes in network conditions or server workloads. In dynamic environments where traffic patterns fluctuate frequently or servers experience varying levels of demand, rule-based algorithms may struggle to dynamically optimize the distribution of incoming requests. This lack of adaptability can lead to inefficient resource utilization, uneven workload distribution, and potential performance bottlenecks on specific servers. Moreover, traditional load balancing methods may overlook factors such as server capacity, processing power, or current load levels when making routing decisions. In scenarios where servers have different capabilities or capacities, rule-based algorithms may not effectively consider these disparities, resulting in imbalanced resource allocation and potential overloading of certain servers while others remain underutilized. Another limitation is the lack of real-time feedback and monitoring capabilities in traditional load balancing approaches. Without mechanisms to collect and analyze real-time data on server performance, network latency, or other critical metrics, rule-based algorithms may struggle to make informed routing decisions in response to changing conditions. This can lead to suboptimal load distribution and reduced overall system efficiency.

DRL holds great potential in the field of load balancing due to its ability to adapt and optimize resource allocation dynamically. In the context of load balancing, DRL leverages a trial-and-error approach to continuously learn and improve decision-making processes based on feedback from the environment. One key advantage of using DRL for load balancing is its capability to handle complex and dynamic environments effectively. Traditional load balancing algorithms may struggle to cope with rapidly changing workloads and network conditions, leading to suboptimal performance. DRL, on the other hand, can adapt quickly to fluctuations in traffic demand and resource availability by learning patterns and making decisions in real-time. Furthermore, DRL offers the advantage of scalability and flexibility in load balancing systems. As the size and complexity of modern networks continue to grow, traditional static approaches may become inefficient or impractical. DRL models can scale to large and diverse environments, making them well-suited for handling the complexities of modern data centers and cloud infrastructures. Another significant benefit of DRL in load balancing is its ability to optimize resource utilization and improve overall system efficiency. By continuously learning from interactions with the environment, DRL agents can intelligently allocate resources based on current demand, thus reducing response times, minimizing latency, and maximizing throughput.

In summary, the use of DRL in load balancing brings about adaptive, scalable, and efficient solutions to the challenges posed by modern networking environments. By leveraging its ability to learn from experience and make data-driven decisions, DRL has the potential to revolutionize the way we manage and optimize resource allocation in complex systems.

In today's distributed computing environment, the application of DRL in the field of load balancing has become crucial for optimizing resource allocation and system performance. By leveraging DRL technology, load balancers can dynamically adapt to changing workloads and efficiently distribute tasks across multiple servers, thereby enhancing system performance and resource utilization. To successfully implement DRL in load balancing, several key aspects need to be carefully considered and designed. Firstly, the design of the state space is essential. In a DRL-based load balancing system, the state space includes the current configuration of the environment, encompassing key metrics such as server loads, task arrival rates, network traffic, queue lengths, and task completion times. Additionally, the state space may also incorporate historical data and patterns that provide valuable insights into system behavior and dynamic workload characteristics. By capturing these important features, the state space enables the DRL agent to make informed decisions based on system states and workload characteristics. Secondly, the design of the action space is critical. Designing the action space involves defining a range of actions that the load balancer can take in a specific state. In the context of load balancing, actions may include selecting servers to process new requests, migrating tasks between servers, or adjusting server capacities. The action space should be carefully designed to empower the DRL agent to effectively manage system dynamics and respond to evolving workloads. When defining the action space, considerations such as server capacities, communication overhead, and task migration costs may need to be taken into account. Next, the design of the reward function is crucial. The reward function is at the core of the DRL agent's learning process, shaping its decision-making behavior and guiding it towards achieving load balancing objectives. A well-designed reward function should align with the overall goals of the load balancing system, such as minimizing response times, maximizing resource utilization efficiency, and preventing server overloads. In formulating an effective reward function, balancing short-term gains with long-term objectives is paramount. Factors such as task completion times, server loads, system performance metrics, and energy consumption may need to be considered to guide the agent towards optimal load balancing outcomes. Selecting an appropriate DRL algorithm is also vital. Common DRL algorithms such as DQN, PG, or Actor-Critic (AC) methods can all be applied in the context of load balancing. The choice of algorithm should consider factors such as the complexity of the load balancing problem, sample efficiency, and learning stability. For instance, if the load balancing problem involves discrete action spaces and can benefit from experience replay, DQN might be a suitable choice. On the other hand, if the problem requires continuous action spaces and policy optimization, PPO or AC methods may be more suitable. The training and evaluation phases involve allowing the DRL agent to learn the optimal load balancing strategy through interaction with a simulated environment or historical data. During the training process, the agent iteratively refines its decision-making strategy to improve load balancing performance. Evaluating the agent's performance using well-defined metrics such as average response times, server loads, system throughput, and other relevant performance indicators is crucial for assessing performance, evaluating the effectiveness of the learned strategy, and identifying areas for improvement.

DRL has shown great potential in optimizing complex systems and decision-making processes. When it comes to load balancing, applying DRL techniques can revolutionize how we manage and distribute workloads across servers efficiently. By leveraging DRL algorithms, systems can learn how to dynamically allocate resources based on real-time demands and feedback. This adaptive approach can lead to improved performance, reduced latency, and better resource utilization in load balancing scenarios. In the future, we can expect DRL to play a significant role in enhancing load balancing algorithms. This technology can adapt to changing network conditions, predict traffic patterns, and optimize resource allocation in ways that traditional methods may struggle to achieve. Overall, the application of DRL in load balancing holds promise for creating more intelligent, efficient, and responsive systems that can handle the increasing demands of modern computing environments.

Chapter 4
Transmission Intelligence

4.1 Learning Channel Encoding

4.1.1 The Conventional Identification Methods and Its Limitation

Blind Recognition of Channel Codes refers to the technique of identifying the type and parameters of the encoding employed from intercepted signals in the absence of prior information. The technology of channel encoding recognition holds potential applications in two main areas: Firstly, for link adaptation in the field of intelligent communication, and secondly, for information warfare under non-cooperative communication conditions.

In the context of link adaptation, the transmitting party needs to select modulation and coding schemes that are suitable for the current electromagnetic environment and channel noise based on the state of the link. The receiving party, in turn, needs to identify the modulation and coding parameters used by the transmitting party based on the received signal, and subsequently perform demodulation and decoding.

In the context of information warfare under non-cooperative communication conditions, the non-cooperative receiving party typically lacks prior knowledge of the channel coding scheme employed by the communicating system on the other end. Therefore, it is necessary to initially recognize and analyze the channel coding before proceeding with the correct reception, decoding, and interpretation of the information.

Traditional methods refer to a category of techniques that concentrate on the structure of indicators or features extracted from the demodulated signal. They aim to identify the coding methods or estimate the coding parameters based on these designed indicators or characteristics. This method has been used since the discovery of blind decoding issues and continues to evolve in response to new requirements. The primary techniques within this category include matrix transformation, rank

H. Zhang and N. Yang, *Deep Learning in Wireless Communications*,
https://doi.org/10.1007/978-981-97-6314-6_4

characteristics, and rank loss analysis, which are fundamental in illustrating differences based on matrix theory principles. Another method is the run feature, which helps differentiate between linear block codes and convolutional codes. Additionally, tools such as original/generating polynomial, generative matrix, and key equations are valuable in turbo codes and convolutional codes, enabling both recognition of coding methods and estimation of parameters. Furthermore, novel matrices like the sparse check matrix for LDPC codes and the information matrix for enhancing anti-noise capabilities are introduced to address specific scenarios.

Traditional methods often have some obstacles in low signal-noise ratio environment, but they help to know more features about different code. That is the reason why people still work on traditional methods. But traditional methods also have limitations as follow:

- When it comes to code identification, most existing recognition approaches are tailored to specific coding methods. Traditional methods fail to provide a universal recognition algorithm due to the fundamentally different mathematical models of linear block codes and convolutional codes. Additionally, for the identification of coding parameters, most traditional methods are designed for specific types of channel coding (e.g., identifying the code length and generator polynomial of BCH codes). There is a lack of a universal recognition algorithm for coding parameters such as code length and code rate, that are applicable to all channel codes.
- In terms of recognition conditions, whether for code type or coding parameters, most traditional recognition algorithms require some prior knowledge and cannot achieve fully blind recognition. Moreover, the performance of some traditional recognition algorithms is limited by the code length and the length of the received signal. As the code length increases and the number of received symbols decreases, the recognition performance of these algorithms deteriorates significantly.

4.1.2 Deep-Learning-Based Blind Recognition of Channel Code

First, we will introduce Primary Techniques in blind recognition of channel. In recent years the research of AI is popular, which encourages scholars to apply AI everywhere including the blind channel decoding.The most commonly used methods in AI include Multi-Modality Features Fusion Network (MMFFN) for space-time block code recognition, CNN for spatial characteristic extraction from sequential data, CNN+RNN for combining CNNs with RNNs to extract deep-seated temporal characteristics, and Deep Residual Shrinkage Network (DRSN) for improving the ACC of recognizing coding methods and parameters, especially in low SNR scenarios.

With the application and update of tools in AI, more and more problems in blind channel decoding will be solved to some extent, so that scholars can extract the characteristics from serial sequences better.

Generally, neural network and deep learning show their ability in blind channel decoding field especially on newly proposed coding method and they are waiting for more use in later research. Then we will introduce the specific application of DRL in blind recognition of channel codes.

Secondly, we will introduce Closed-set identification of channel encoding. In the field of channel coding, closed-set recognition tasks are typically conducted under the common closed-set assumption. This assumes a defined set of candidates for recognition, with training and testing data both originating from the label space of this candidate set. The goal of this problem is to improve recognition performance and reduce misclassification rates. Such recognition tasks hold significant importance in the context of Adaptive Modulation and Coding (AMC) technology.

To solve this problem, a closed-set recognition algorithm called CCR-Net is proposed. This algorithm utilizes specially designed convolutional shrinkage blocks to construct a feature embedding model, extracting encoding features through nonlinear transformation layers and attention mechanisms. Additionally, the introduction of center loss technique helps to minimize the distance between features of the same class and enlarge the distance between features of different classes, thereby improving recognition performance. Experimental results demonstrate that CCR-Net performs well in low SNR conditions.

Finally, we will introduce open-set identification of channel encoding. Open-set recognition refers to the task of identifying unknown encoding categories and rejecting their recognition based on the analysis of intercepted signals. The goal of this problem is to improve the detection ACC of unknown encoding categories and reduce misclassification rates. In contrast to closed set recognition, the recognition task is more challenging.

To solve this problem, an open-set recognition algorithm called CCR2CNN is proposed. This algorithm utilizes a multi-task learning framework to simultaneously perform closed-set encoding classification and signal reconstruction tasks. By comparing the difference between the reconstructed signal and the original signal, the detection of unknown encoding categories is achieved. Additionally, the extreme value theory model is introduced to simulate the distribution of reconstruction errors, further improving the open-set recognition performance. Experimental results demonstrate that CCR2CNN can accurately reject unknown encoding test samples while recognizing known encoding categories, showing good open-set recognition performance.

In conclusion, there are two main kinds of method in blind channel decoding. The first is to distinguish tradition informatics characteristics of code-word sequence, which usually performs well in the appropriate coding, but most of these are poor in versatility. The Second and also the new one is to recognize the method and parameters of channel coding by neural network and deep learning. As a common disadvantage, this kind of methods need a lot of calculation to train the model and people cannot explain the logic of operation, but they also perform well and own good transferability for similar channel coding methods. That is important for the blind decoding of newly proposed encoding methods.

In future, the research of tradition methods is necessary but the best way is to use neural network and deep learning along with the original scheme, that will improve the explain ability, versatility and performance in the same time.

4.2 Channel Estimation and Equalization Methods for Large-Scale MIMO-OFDM Systems

4.2.1 Channel Estimation Methods

Large-scale MIMO systems, due to the significantly larger number of antennas deployed at the base Station compared to traditional MIMO systems, can greatly increase system capacity and possess high spectral and energy efficiency. Therefore, they are considered as one of the key technologies in 5G wireless communication systems. Orthogonal OFDM technology divides the channel bandwidth into numerous orthogonal subcarriers, allowing each subcarrier to occupy a small bandwidth for low-speed data transmission. The combination of large-scale MIMO and OFDM techniques in large-scale MIMO-OFDM systems leverages the advantages of both technologies to effectively combat frequency-selective fading, increase channel capacity, and enhance spectral efficiency.

Channel estimation is extremely important for the performance of wireless communication systems. The transmitted signal often experiences distortion due to the multipath channel characteristics, so it is essential to accurately estimate the channel impulse response at the receiver to recover the transmitted signal. In channel estimation theory, there are three commonly used algorithm types: blind channel estimation, semi-blind channel estimation, and non-blind channel estimation.

Blind channel estimation refers to the estimation of channel state information without the need for transmitting training sequences or pilot symbols. It relies on extracting the structure and inherent properties of the received signal to estimate the channel. This approach reduces the waste of additional spectral resources as it doesn't require transmitting known information. However, blind channel estimation algorithms require a large number of OFDM symbols to obtain reliable cyclic correlation estimation for channel estimation. These algorithms have high estimation complexity, longer processing times, often do not achieve good bit error rate performance, and their most significant drawback is the inability to operate when digital modulation techniques and coding schemes are unknown. Non-blind channel estimation methods involve tracking and adjusting the parameters to be estimated gradually, based on the estimation criteria and with the help of training sequences or pilot symbols. The objective is to determine the estimated values of each parameter. On the other hand, semi-blind channel estimation combines the advantages of high spectral efficiency in blind channel estimation and relatively better performance, lower computational complexity in non-blind channel estimation. It is a compromise channel estimation method that achieves good recovery performance by transmitting only a small amount of known data from the transmitter, to some extent mitigating the pilot contamination problem.

In large-scale MIMO-OFDM systems, the ACC of channel estimation directly affects the bit error rate and mean-square error performance of the system. While obtaining a well-performing channel estimation algorithm is desirable, it inevitably leads to increased algorithm complexity. The level of complexity required to implement the method is also a critical factor in determining whether an estimation algorithm is suitable for practical systems. In the fundamental theory of wireless communication, classical channel estimation algorithms include the least squares (LS) algorithm, MMSE algorithm, Linear Minimum Mean Square Error (LMMSE) algorithm, etc.

The LS Algorithm: The LS algorithm minimizes the sum of squared channel estimation errors as its criterion. The ideal LS algorithm performance is often used as a reference to evaluate the effectiveness of proposed channel estimation methods. Due to its simplicity and the fact that it doesn't require any channel statistical information, the LS algorithm is widely applied in channel estimation methods. However, it disregards noise interference, resulting in poor performance in complex large-scale MIMO-OFDM system communication environments.

The MMSE Algorithm: To overcome the drawback of LS algorithm being sensitive to noise, the MMSE algorithm can be used for channel estimation when the statistical characteristics of the channel and interference information are known. The MMSE algorithm relies on the LS algorithm but has much higher complexity since it requires prior knowledge of the statistical properties of the channel, including channel auto-correlation matrix and noise variance. Obtaining such information in practical wireless communication systems is challenging. Therefore, while the MMSE algorithm considers performance to the extreme, it is difficult to apply in practical scenarios.

The LMMSE Algorithm: The LMMSE algorithm is an improvement over the MMSE algorithm. It reduces the complexity of the estimator by averaging the transmitted data when stationary random signals are transmitted over wireless channels. The LMMSE algorithm has higher computational efficiency and better performance compared to the LS algorithm, making it more suitable for practical systems. However, in large-scale MIMO-OFDM systems, there are still challenges in dealing with the enormous amount of data and solving the related problems of solving auto-correlation matrices and matrix inversions.

At the same time, DNN algorithms offer superior adaptive capability, nonlinear modeling ability, and large-scale data processing capability in channel estimation for large-scale MIMO-OFDM systems. They provide a balance between model complexity and ACC and can directly learn the mapping relationship for channel estimation from raw data through end-to-end learning. This makes DNNs a promising and powerful tool for channel estimation in large-scale MIMO-OFDM systems.

Neural network-based channel estimation algorithms for large-scale MIMO-OFDM systems can generally be classified into four types. Firstly, channel parameters are initially estimated using the LS criterion, followed by the utilization of a neural network for principal component analysis under blind criteria constraints to enhance the estimation performance. Secondly, NNs are employed to obtain initial channel values, particularly in fast-fading channels, with subsequent filtering algorithms utilized to track variations in the channel information. Thirdly, in pilot-based channel estimation methods, neural network-based approaches replace interpolation filters to acquire channel information at non-pilot positions. Finally, NNs are considered as nonlinear models to represent the mathematical relationship between the channel impulse response and subcarrier position indices.

Next, we will provide a detailed introduction to the DNN-based channel estimation method for large-scale MIMO-OFDM systems. The objective of this algorithm is to minimize the difference between the network's output and the desired target, aiming for the network's output to be as close as possible to the expected result. Therefore, it requires minimizing the cost function, which is formulated as follows:

$$E_i = \frac{1}{2} \left\| \boldsymbol{a}^L - \boldsymbol{o} \right\|^2 = \frac{1}{2} \left\| f \left(\boldsymbol{W}^L \boldsymbol{a}^{L-1} + \boldsymbol{b}^L \right) - \boldsymbol{o} \right\|^2 \tag{4.1}$$

where, W^L represents the connection weight values from layer $(L-1)$ to layer L, $a^{(L-1)}$ denotes the output matrix of layer $(L-1)$, b^L represents the bias matrix of layer L, and o represents the target output set of the network.

The iterative algorithm initializes by utilizing the LS estimator with the FFT method for OFDM signal reconstruction at the signal receiver to obtain the channel frequency response estimation. Subsequently, training data containing subcarrier position indices and channel estimation values obtained through the LS method for known pilot positions are inputted into the DNN network. The DNN model's output layer computes the output a^L, partial cost function value, and derivatives with respect to network parameters. A determination is made based on whether the cost function converges: if so, parameters are saved; otherwise, parameters are updated, and outputs are recalculated. This process continues until reaching a predefined number of iterations or when the error between consecutive iterations becomes negligible. Finally, the optimal weights and biases of the trained DNN model are returned. If data subcarrier position indices are inputted, the DNN network outputs the desired channel frequency response for the data subcarrier positions (Fig. 4.1).

4.2.2 Channel Equalization Methods

The channel equalization problem in large-scale MIMO-OFDM systems is also a topic worthy of in-depth exploration. Channel equalization refers to the process of using equalization techniques at the receiver end of the system, after the FFT is performed, to restore the transmitted signals from the transmitter and reduce or eliminate interference and distortion caused by the channel.

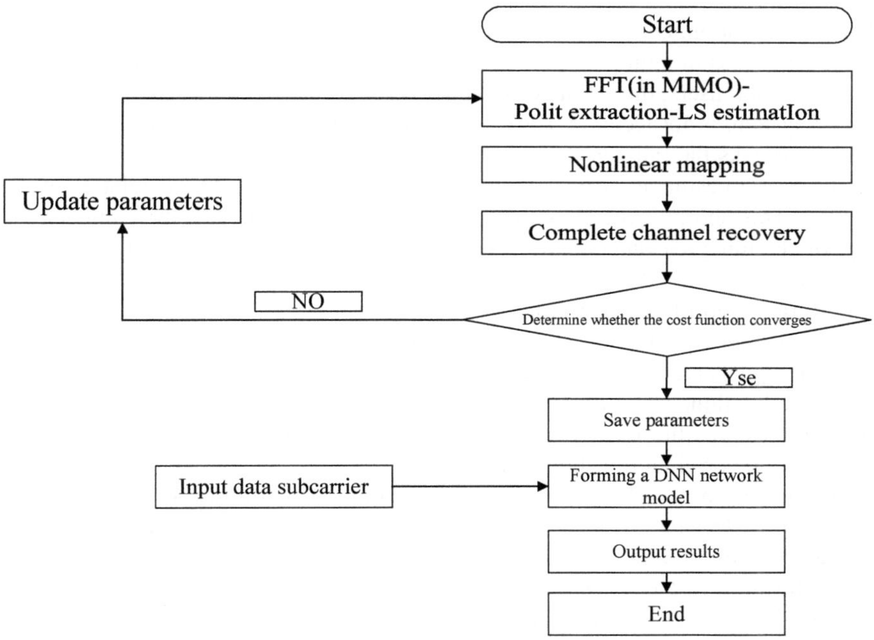

Fig. 4.1 The iterative process of channel estimation algorithm for large-scale MIMO systems based on DNN network

There are two main categories of channel equalization methods for large-scale MIMO-OFDM systems: linear equalization and nonlinear equalization. Linear equalization methods include zero forcing (ZF) and MMSE equalization. Linear equalization methods have the advantage of lower computational complexity, but they often have poor detection performance, leading to the abandonment of these methods in many wireless communication systems that prioritize communication quality. However, in certain specific scenarios, such as the SC-FDMA technology used in the LTE uplink, these linear equalization methods still have research significance. Nonlinear equalization methods compensate for the limitations of linear equalization methods in wireless multipath channels and include maximum likelihood (ML) equalization, QRM-MLD, LR algorithm, etc.

Traditional equalizers perform well in quasi-static or slow fading channel conditions but perform poorly in time-varying channels. In practical communication systems, such as mobile communication, which belong to time-varying fast fading channels, researchers have attempted to use dynamic NNs as equalizer structures instead of conventional static equalizers.

Also, in large-scale MIMO-OFDM systems, traditional non-neural network-based channel equalization algorithms require accurate channel state information at the receiver, which is the channel estimation technique. The performance of equalization detection in large-scale MIMO-OFDM systems largely depends on the ACC of channel estimation. Due to the mutual interference of antenna arrays at the base station, energy loss in multipath propagation, and artificial interference from multiple small cells and users, the estimated channel state information in large-scale MIMO-OFDM systems is not accurate. The use of neural network-based channel equalization methods breaks the inherent transmission model structure of the system, eliminating the need to estimate channel state information after signal demodulation at the receiver. This shortens the signal processing flow and eliminates the cumbersome steps of obtaining relevant parameters, allowing for the direct recovery of the original transmitted data from the received data. The specific description of the DNN channel equalization algorithm is as follows:

The channel equalization problem is treated as a classification problem, where the input signals are mapped to different positions in a constellation diagram using digital modulation. The different positions represent different categories. The DNN equalizer restores the received signals to their corresponding categories, thereby recovering the original transmitted signals.

The entire process of the equalization algorithm is as follows: For a large-scale MIMO-OFDM system with $N_t \times N_r$ antennas, the DNN equalizer uses known pilot data as the network input. The received pilot data, after removing the cyclic prefix and performing fast Fourier transform, is used as the network output. The DNN equalizer is trained by minimizing the error between the input and output data. After training, the input weights and hidden layer biases are reset using the gradient descent algorithm. If the error does not meet the requirements, the training continues.

The network output data from the DNN equalizer is classified into the correct categories using the maximum pooling decision rule, and the output neurons in the output layer are mapped to the corresponding digital modulation symbols. The DNN equalizer receives different categories of original transmitted signals at its receiving end and calculates the bit error rate between the received and transmitted data. Since accurate channel state information is not required, the bit error rate is the only performance metric.

In conclusion, with the rapid development of deep learning and neural network technologies, their fundamental ideas have been introduced into various levels of communication systems. By using artificial intelligence techniques to solve complex communication problems, the performance of communication algorithms can be significantly improved while reducing the algorithm's execution time. Communication systems based on neural network methods adjust network parameters by minimizing the error between known pilot data and network outputs, optimizing all processing modules in the communication system, and breaking the modular structure to achieve the overall optimal performance.

In terms of future research directions, it is worth considering more complex channels and the applicability of this method in specific communication environments. The next focus of work will still be on reducing the computational complexity of the algorithm, improving the real-time performance of the network, and further enhancing the estimation ACC of the network. Also, for fast-varying channels in practical communication systems, exploring more flexible and adaptive channel estimation and equalization methods.

4.3 Deep Clustering Models

In the era of big data, with the increasing availability of data and the growing complexity of data structures, traditional clustering algorithms are becoming inadequate to meet the demands of applications such as data analysis and computer vision. DL, on the other hand, has the ability to handle complex data structures, redundant information, and large-scale data, making it a natural choice to address the challenges posed by the big data era. By incorporating deep learning into unsupervised clustering, we can leverage its power to handle complex data structures and overcome the difficulties associated with big data.

The essence of deep clustering is to learn a clustering-guided feature representation using NNs and fit the inherent clustering patterns in the data. Therefore, deep clustering models have three commonly used evaluation metrics: Unsupervised Clustering ACC, Normalized Mutual Information (NMI), and Homogeneity. ACC is used to assess clustering ACC and typically requires the number of clusters to be equal to the number of true classes. It aims to find the mapping that maximizes ACC among all possible mappings. NMI measures the difference between the predicted label distribution and the true label distribution. Homogeneity evaluates the homogeneity of clustering clusters, which refers to the likelihood that samples assigned to the same cluster belong to the same category.

The two key factors of deep clustering models are the neural network and the clustering patterns. By examining the loss functions of existing deep clustering models based on these key factors, the overall loss function can be unified as follows:

$$\mathrm{L} = \alpha \mathrm{Lc} + \beta \mathrm{Ln}. \tag{4.2}$$

According to this formula that could see the complete loss function consists of two parts: the network loss L_n, which is used to learn a feature representation conducive to clustering, and the clustering loss L_c, which is used to fit the specified clustering patterns. From the perspective of NNs, deep clustering models can be categorized into those based on feed-forward NNs, those based on autoencoders, hose based on generative models, and those based on graph CNNs.

4.3.1 Clustering Models Based on Feedforward NNs

First, we will introduce clustering models based on feedforward NNs. The most famous model in feedforward NNs is the convolutional neural network. Feedforward NNs generally require sample labels as supervised signals, which limits their application in unsupervised clustering. Overcoming this limitation is one of the challenges for deep clustering models based on feedforward NNs. The Recurrent Framework Agglomerative Deep Clustering (RFADC) model is a deep clustering model built on CNNs. Feedforward NNs typically require sample labels as supervised signals.

The RFADC model constructs a recurrent framework based on agglomerative clustering. In each iteration, clusters from the previous iteration are merged based on their similarity, and the cluster labels from the previous iteration are used to update the feedforward convolutional neural network for feature extraction. The underlying idea behind the RFADC recurrent framework is that good clustering results can extract feature representations conducive to clustering, and clustering-guided feature representations can enhance clustering performance, forming a mutually reinforcing loop structure.

4.3.2 Deep Clustering Models Based on Autoencoders

Next, we will introduce deep clustering models based on autoencoders. Autoencoders are the most famous algorithms in the field of unsupervised feature learning, consisting of an encoder and a decoder. The encoder maps the input sample x to a latent feature h. The decoder reconstructs the original sample x from the latent feature h, aiming to make the reconstructed sample r approximate the original sample h. Autoencoders can generate features that represent the essence of the samples when reconstructing the original samples. Since one of the key factors in deep clustering is learning a feature representation that is conducive to clustering, autoencoders can be a preferred choice for deep clustering. The reconstruction loss of autoencoders can generally be regarded as the network loss L_n of the deep clustering model, and the clustering rules designed from different perspectives can be regarded as the clustering loss L_c of the deep clustering model.

Typical Deep Clustering Models

Below, we will introduce some typical deep clustering models based on autoencoders. First, we will introduce the Deep Clustering Network (DCN). DCN model is the most classic model in autoencoder-based clustering algorithms, combining autoencoders with k-means algorithm. The objective function is as follows:

$$\min_{W,Z,M,\{S\}} \sum_{i=1}^{N} \left(l\left\{g\left[f\left(x_i\right)\right], x_i\right\} + \frac{\lambda}{2} \left\| f\left(x_i\right) - \boldsymbol{MS} \right\|_i \|^2 \right), \tag{4.3}$$

where $f()$ represents the encoder, $g()$ represents the decoder, $l()$ represents the reconstruction error function, M is the center matrix composed of the center vectors of each cluster, and S_i is the one-hot vector indicating the cluster membership of sample i. The objective function of this model includes the center matrix and the allocation vectors, and the optimization of the entire objective function requires the use of a custom alternating optimization algorithm.

Deep Embedding Network (DEN)

Secondly, we will introduce the DEN. DEN is another autoencoder-based clustering model that adds two constraint terms on top of the autoencoder to learn a feature representation that is conducive to clustering. Finally, the k-means algorithm is applied on this clustering-friendly feature to provide clustering results.

Deep Embedded Clustering (DEC)

Third, we will introduce DEC. DEC is one of the most representative methods in the field of deep clustering. Its introduction has attracted widespread attention in the deep learning community. The DEC model pretrains an autoencoder, selecting the encoder part for feature extraction. It uses a soft distribution to predict cluster labels for the samples based on the extracted features, and a hard distribution to select the cluster labels with high confidence. Minimizing the KL divergence between the soft distribution and the hard distribution gradually guides the clustering results towards high-confidence distributions. The soft distribution converts the similarity between samples and cluster centers into a probability distribution. The hard distribution strengthens the soft distribution, enhancing the cluster assignments with high confidence and weakening those with low confidence. This strengthening operation is achieved by squaring and normalizing the soft distribution.

Deep Subspace Clustering Networks (DSC-Nets)

Finally, we will introduce DSC-Nets. DSC-Nets are a type of clustering method based on deep learning, aiming to cluster data samples into different subspaces. DSC-Nets typically consist of two main components: an encoder and a clustering layer. The encoder maps the input data to a low-dimensional latent space representation, performing feature extraction and abstraction through multiple hidden layers. The clustering layer then performs clustering operations based on the feature vectors output by the encoder, assigning samples to different clusters.

The training process of DSC-Nets is typically an end-to-end process, optimizing network parameters by minimizing a clustering loss function. The clustering loss function usually consists of two parts: a reconstruction loss and a clustering loss. The reconstruction loss measures the difference between the original data and the reconstructed data, encouraging the network to learn better feature representations. The clustering loss measures the consistency between the clustering results and the true labels, driving the network to correctly cluster samples into the corresponding subspaces.

In conclusion, it is effective in handling high-dimensional data and complex data structures, improving clustering ACC and stability. However, the DSC-Nets model has high space and time consumption, making it unsuitable for large-scale datasets.

4.3.3 Deep Clustering Models Based on Generative Models

Then, we will introduce deep clustering models based on generative models. The purpose of introducing generative models into the clustering domain is to leverage their ability to capture data distributions and capture the posterior probability distributions of data samples belonging to different clusters. The most well-known generative models are Variational Autoencoders (VAE) and Generative Adversarial Networks (GAN). By incorporating these generative models, we can develop deep clustering models based on VAE and GAN.

4.3.3.1 Deep Clustering Models Based on VAE

First, we will introduce deep clustering models based on VAE. VAE is a variant of the autoencoder in the context of generative models. When introducing VAE into the clustering domain, the assumption of following a standard Gaussian distribution is changed to following a Gaussian mixture distribution composed of multivariate Gaussian distributions. The existing deep clustering models, such as Gaussian Mixture VAE (GMVAE) and Variational Deep Embedding (VaDE), are based on this new assumption. These models can not only output the probabilities of samples belonging to different clusters but also generate samples of specified categories. However, the computational complexity of these models is high.

4.3.3.2 Deep Clustering Models Based on GAN

Secondly, we will introduce deep clustering models based on GAN deep clustering models based on GAN aim to capture the data distribution by maximizing and minimizing the adversarial training between the discriminator D and the generator G. The generator G generates a sample from the latent features z, which come from the prior distribution $p(z)$. The discriminator D is trained to distinguish between real samples and fake samples generated by G. These deep clustering models based on GAN have the ability to capture the posterior probability distribution of data samples belonging to different clusters. Some models are specifically developed for clustering tasks, while others treat clustering as an application of the model. Examples of such models include Deep Adversarial Clustering (DAC), Categorial Generative Adversarial Network (CatGAN), and Information Maximizing Generative Adversarial Network (InfoGAN).

4.3.4 Deep Clustering Model Based on Graph Neural Networks (GNNs)

At last, we will introduce the deep clustering model based on graph neural networks. Graph convolution operations can handle non-aligned data, such as graph-structured data, and the extracted feature information contains the structural information of data nodes. Therefore, the advantage of introducing graph CNNs into unsupervised clustering is that it can extract feature representations that are conducive to clustering. Graph convolution operations require the dataset to have graph structure information, while the datasets that need clustering generally do not have graph structure.

To address this issue, the Adversarial Graph Auto-Encoders (AGAE) model integrates clustering to construct a consistency graph to introduce graph convolution operations into the clustering domain. The AGAE model is a deep clustering model based on graph neural networks. It introduces graph convolution operations into the clustering domain by constructing a consensus graph through the integration of clustering. The specific construction method of the model involves using multiple traditional clustering methods to cluster the dataset and obtain multiple clustering assignments. These assignments are then used to construct a joint matrix as the graph structure of the dataset. The AGAE model utilizes graph convolution operations to extract the structural information of the data nodes and optimizes the clustering results through adversarial training. This model can extract feature representations that are conducive to clustering and is suitable for handling non-aligned data, such as graph-structured data.

Based on the analysis of the aforementioned deep clustering models and research on the structure of deep clustering, the future focus of deep clustering can be summarized as follows: Clustering theory research is essential to explore the theoretical basis of deep clustering, providing theoretical guidance for further research. Deep clustering models leverage different NNs to enhance clustering performance, but the suitability of NNs varies depending on the scenario. For instance, while CNNs are suitable for image datasets, RNNs are more appropriate for sequence datasets. Most existing deep clustering models are developed for image datasets, so exploring the application of RNNs in unsupervised clustering for text sequence datasets is a future task. Another key focus is balancing the clustering constraints, as the clustering performance of deep clustering models results from the combined effects of multiple constraints, each with varying levels of importance. Therefore, it's crucial to balance these constraints effectively.

4.4 Adaptive Content Caching

4.4.1 The Motivation of DRL for Adaptive Content Caching

Given the substantial surge in data traffic across both wired and wireless communication channels, the evolution of next-generation networks becomes crucial. This evolution encompasses future Internet architectures, content delivery infrastructure, and cellular networks, all of which necessitate cutting-edge technologies to address the ever-growing demand for data. Identified as an attractive solution is caching, which involves the storage of reusable content in geographically distributed storage-enabled network entities, enabling faster retrieval during subsequent requests. The underlying principle is that the adverse impacts of peak traffic periods can be mitigated by proactively storing "anticipated" highly popular content in these storage devices, especially during off-peak periods. Caching popular content is envisioned to yield substantial savings in terms of energy consumption, bandwidth utilization, and overall costs, while simultaneously enhancing user satisfaction.

The reasons for using DRL in adaptive content caching are as follows:

First, DRL has excellent dynamic adaptability. Adaptive content caching requires dynamically selecting cached content based on user demands and network conditions. Traditional caching strategies are often static and cannot adapt to changing environments. DRL, on the other hand, can learn the optimal strategy through interaction with the environment, allowing for dynamic adjustment of caching policies based on real-time user behavior and network status. This enables the cache to better adapt to changing demands and environments.

Second, DRL has an advantage in handling complex data. Adaptive content caching involves processing large amounts of complex data, including user requests, network status, and content features. Traditional caching strategies often struggle to handle such complex data and fail to fully utilize the information to make optimal caching decisions. DRL, with its ability to utilize DNN, can effectively process complex data, extract useful features, and learn to understand user demands and content features, leading to more accurate caching decisions.

Third, DRL has the advantage of RL, which allows for optimizing strategies through learning. DRL is a form of RL that learns the optimal policy through interaction with the environment. In adaptive content caching, DRL can learn the optimal caching policy by interacting with user requests and network conditions, continuously optimizing caching decision performance. Compared to traditional caching strategies, DRL can better adapt to different scenarios and demands, providing more efficient caching services.

In summary, the reasons for using DRL in adaptive content caching lie in its dynamic adaptability, ability to handle complex data, and the advantages of RL. These characteristics make DRL an effective method for optimizing the performance of adaptive content caching.

To implement Adaptive Content Caching, sophisticated algorithms, and machine learning techniques are often employed to analyze and process the vast amount of data involved. These algorithms continuously monitor network conditions, track content popularity, and adapt the caching strategy in real-time to optimize content delivery. Adaptive Content Caching aims to improve network efficiency, reduce latency, and enhance the user experience by dynamically adapting the caching strategy based on various contextual factors.

4.4.2 The Model and Problem Statement of Adaptive Content Caching with DRL

In this chapter we will introduce a problem model on adaptive caching in hierarchical content delivery networks. The objective is to efficiently utilize the limited storage capacity of network entities by caching popular content during off-peak periods. This can enhance the performance of the network infrastructure and improve the user experience during peak periods.

The model considers a network consisting of a parent node and N leaf nodes. The parent node is connected to the cloud through a back haul link. Each node stores files to respond to file requests. The leaf nodes serve the end users connected to them, providing the requested content locally if available, otherwise obtaining it from the parent node. The parent node observes the aggregate requests from the large number of users served by the N leaf nodes.

The problem statement is how to effectively manage the caching nodes in this network to minimize the overall cost. Specifically, the problem is to find an optimal strategy that maximizes the satisfaction of user file requests within given resource constraints and minimizes the cost of fetching files. This is an optimization problem with the objective of finding an optimal strategy that minimizes the overall cost on a long-term cumulative time scale.

Also, the problem model assumes that the popularity of file requests varies over time and that there is spatial and temporal evolution of file requests among different leaf nodes. This means that file requests between different leaf nodes may have different levels of popularity and rates of evolution.

The challenge in the problem model is how to effectively manage caching nodes under dynamic evolution of file requests and network topology conditions. Traditional caching policies such as Least Recently Used (LRU) and Least Frequently Used (LFU) are unable to cope with the dynamic changes in file popularity and network topology. Therefore, there is a need to develop a caching policy that can adapt to the local policies of leaf nodes and the spatial and temporal evolution of file requests.

To solve this problem, a two-timescale approach is adopted. On the fast timescale, the file requests received by the leaf nodes exhibit rapid temporal evolution, requiring quick decision-making. On the slow timescale, the parent node observes aggregate requests that change slowly. Based on this observation, an adaptive RL approach is proposed to learn a policy function through interaction with the environment, aiming to minimize the long-term cumulative cost.

4.4.3 DRL for Adaptive Content Caching

In this part, we will introduce the basic process of DRL based adaptive caching algorithm. The goal of this algorithm is to adapt to dynamic file requests and leaf node caching policies through interaction with the environment and minimize the overall cost over a long-term cumulative time scale.

To address the complexity of caching decisions, an RL framework is proposed. RL allows the parent node to learn and make optimal caching decisions through interaction with the leaf nodes. This approach utilizes a DQN, which is a DNN that approximates the Q-values of different caching actions.

In this problem model, a major challenge is the large continuous state space. To overcome this issue, a scalable deep RL approach is adopted. This method learns from data using DNN and provides compact representations of high-dimensional states.

The algorithm follows a sequence of steps: First, it initializes network parameters and the cache state, setting up the DNN with the appropriate number of nodes in each layer and initializing the cache capacity and content. Next, it interacts with the environment, selecting actions (caching a file, replacing a file, or taking no action) based on the current cache state and file requests to update the cache policy. The cache policy is then updated accordingly: files are added to the cache, replaced, or left unchanged. Subsequently, the reward is calculated based on the updated cache policy and file request satisfaction. The DNN parameters are updated using the reward value to enhance the cache policy's performance, employing optimization algorithms like gradient descent. This process iterates until a predetermined number of training iterations or a stopping condition is met.

By continuously interacting with the environment and updating the parameters, the DRL for adaptive caching algorithm can gradually learn the optimal cache policy to maximize user satisfaction and minimize the overall cost. In summary, DRL for adaptive caching is a DRL-based adaptive caching algorithm. Through interaction with the environment and parameter updates, the algorithm can gradually learn the optimal cache policy to maximize user satisfaction and minimize the overall cost.

4.5 DRL for Computing Offloading in Mobile Edge Computing

4.5.1 The Motivation of DRL for Computing Offloading in Mobile Edge Computing

Mobile-edge computing (MEC) has emerged as a promising computing paradigm in the 5G architecture, which can move computation, caching, and network functions toward the network edges. It could also empower user equipments (UEs) with computation and energy resources offered by migrating workloads from UEs to the nearby MEC servers.

Mobile Computing Offloading (MCO) constitutes a pivotal process within MEC. MCO represents a promising approach whereby resource-intensive tasks, or at least a portion thereof, are offloaded to the resource-rich servers in proximity to the MEC, thereby alleviating constraints on client devices.

Although the issues of computation offloading and resource allocation in MEC have been studied with different optimization objectives, they mainly focus on facilitating the performance in the quasistatic system, and seldomly consider time-varying system conditions in the time domain. The emergence of deep learning as a robust tool for processing massive datasets has captured the attention of researchers, as it enables the extraction of real-world data in complex and noisy environments.

Edge computing encounters diverse resource allocation challenges across different layers, including CPU cycle frequency, access coverage, RF, and bandwidth. Consequently, it necessitates a repertoire of robust optimization tools to enhance system efficiency. And in the absence of any a priority knowledge, DRL can intelligently enhance edge networks to adeptly capture the latent dynamics of the environment, thereby learning policies to achieve optimal long-term objectives through iterative interactions within specific contexts. This characteristic endows DRL with a distinctive potential when designing computational offloading and resource allocation schemes in dynamic systems.

4.5.2 The Model of Computation Offloading

Due to the constraints imposed by battery power, the energy consumption of terminal devices is of paramount importance. Simultaneously, the processing time delay directly impacts user experience. Presently, the majority of research employs a weighting factor to balance these two objectives, normalizing multiple goals into a single optimization target. However, the normalization of objectives leads to the loss of some distinguishing features, and the determination of the correct weighting factor is challenging. Therefore, researchers are now considering the optimization objectives as the time delay and energy consumption. This approach models the computational offloading problem in mobile edge networks as a multi-objective optimization problem, with the parameters requiring optimization including offloading decisions, device CPU operating frequency, and transmission power.

For the purpose of analytical convenience, in line with the approach adopted in numerous literature on computation offloading, the entire problem scenario is typically regarded as a quasi-static field, wherein, over a period, the number of mobile devices, the data for each computational task, and the channel state of the wireless link remain fixed.

According to the overview of computation offloading, it is evident that the processing of a computational task can be accomplished through local execution or offloading to an edge server. For the sake of convenience, the model assumes that the tasks are indivisible, meaning that the computational task of each device is either entirely locally executed or entirely offloaded for remote processing by the server. Let S_i represent the offloading decision for the i-th device; if the device offloads the task to the MEC server for execution, $S_i = 1$, otherwise, the device undertakes local processing, in which case $S_i = 0$.

Thus, the computational cost of the i-th terminal device's computing task can be expressed as:

$$\min G_i = [G_i^T, G_i^P] \tag{4.4}$$

$$G_i^T = s_i t_i^c + (1 - s_i) t_i^L \tag{4.5}$$

$$G_i^P = s_i e_i^c + (1 - s_i) e_i^L. \tag{4.6}$$

In the above equation, G_i^T and G_i^P respectively denote the computational latency and energy consumption of task τ_i. And the symbol t_i^C in the formula represents the time consumed when offloading the task to the MEC server, where C denotes the execution of the task offloading to the edge server, t_i^L denotes the time consumption for processing the task locally, e_i^C represents the transmission energy consumption of the mobile terminal device, and e_i^L signifies the dynamic power consumption.

In conclusion, the computation offloading problem in MEC networks is modeled as a multi-objective optimization problem. Specifically, the set of all mobile devices is denoted as $N = \{1, 2, \ldots, N\}$, and the set $s = \{s_1, s_2, \ldots, s_N\}$ represents the task offloading decisions. The CPU operating frequencies of all devices when performing local computation are denoted as $f = \{f_1, f_2, \ldots, f_N\}$, and the transmission powers of all devices are denoted as $p = \{p_1, p_2, \ldots, p_N\}$. The parameters s, f, and p are to be optimized, and the entire optimization problem can be formulated as follows:

$$\min_{(s,p,f)} G = [\sum_{i=1}^{N} G_i^T, \sum_{i=1}^{N} G_i^P] \tag{4.7}$$

$$C1 : s_i t_i + (1 - s_i) t_i^L \leq T_i^{\max} i \in N \tag{4.8}$$

$$C2 : f_{\min}^L \leq f_i^l \leq f_{\max}^L i \in N \tag{4.9}$$

$$C3 : 0 \leq p_i \leq p_{\max} i \in N \tag{4.10}$$

$$C4 : s_i \in \{0, 1\} i \in N. \tag{4.11}$$

The constraint condition $C1$ signifies that the total time consumption of each task must not exceed the maximum tolerable delay t_i^{Max}. $C2$ indicates that the CPU frequency of the mobile device can only be dynamically adjusted within the permitted range. $C3$ represents that the maximum transmission power of the device is p_i^{Max}. $C4$ specifies that the offloading of tasks is a binary process, wherein the task's processing method is limited to local execution or offloading, with no possibility of task fragmentation.

4.5.3 The Algorithmic Procedure of DRL

The optimization problem described above presents significant complexity due to its non-convex nature, making traditional solution methods impractical. As a result, researchers have increasingly turned to modeling the computation offloading problem within MEC systems as a MDP. This approach enables the learning of effective strategies, even in intricate environments. Moreover, DRL has been utilized to tackle the challenges related to computation offloading and resource allocation in this context.

MDP encompasses three pivotal elements: states, actions, and rewards. In conjunction with the preceding optimization problem, the following definitions are established:

- State Space: the state S $\in S$ encapsulates the state information of the MEC system and the task information offloaded to the server. Thus, the state at time slot t can be represented as:

$$S(t) = \{Q_{MD}(t), s(t)\} \in S \tag{4.12}$$

  Here, $Q_{MD}(t)$ denotes the state information of the mobile device, encompassing CPU operating frequency and transmission power.
- Action space: $a_i \in S$ represents the actions that the device can execute, including unloading decisions, local execution with CPU operating frequency, and emission power during unloading processing. Thus, the action at time slot t can be expressed as:

$$a(t) = \{s(t), f(t), p(t)\} \in A \tag{4.13}$$

- Reward function: the setting of rewards should be linked to the optimization objectives and should also take into account the system's constraints. Therefore, the reward at time slot t can be expressed as:

$$r(t) = \begin{cases} -G(t), & \forall i \in 1, 2, \ldots, n, \quad C_i \text{ is satisfied} \\ -10, & \exists i \in 1, 2, \ldots, n, \quad \neg C_i \text{ is satisfied} \end{cases} \tag{4.14}$$

In pursuit of minimizing the overall cost of the system, the reward is set as the negative of the objective function when the constraints are met. In cases where the constraints are not satisfied, a substantial penalty is assigned to constrain the behavior of the intelligent agent.

In addressing the aforementioned model, RL is broadly categorized into three approaches: value-based methods, policy-based methods, and AC methods. The value-based approach emphasizes the attainment of maximal value, essentially selecting actions that yield the highest value to ultimately derive the optimal strategy. Value-based methods are suitable for scenarios with discrete action spaces. Representative algorithms include Q-Learning and State-Action-Reward-State-Action(SARSA) algorithms.

The policy-based approach serves as a complement to the value-based methods, emphasizing action policies and aiming to learn the policy function that directly maps the current state of the agent to the corresponding action. The objective of this approach is to find an optimal policy that enables the agent to attain the maximum cumulative reward in a specific task. Policy-based methods are suitable for scenarios with continuous action spaces. Representative algorithms include the PG algorithm.

On the other hand, AC methods integrate both value and action considerations. Currently, the most prevalent algorithms adopt this approach, with representative examples being the AC algorithm and the Asynchronous Advantage Actor-Critic algorithm.

4.6 Beamforming with DRL in MIMO

4.6.1 The Motivation of DRL for Beamforming in MIMO

Future 5G/6G wireless networks will be increasingly using millimeter waves (mm Waves), where fast and efficient beamforming is vital for providing continuous service to highly mobile devices in the presence of interference and signal attenuation, manifested by blockage. A crucial facilitator for establishing connectivity between swiftly moving UE and at least one AP at any given moment involves the timely and efficient implementation of interference mitigation and beam steering (collectively referred to as beamforming).

Fully digital beamforming is associated with significant costs, power consumption, and the necessity for complex hardware. Conversely, hybrid beamforming has the potential to deliver comparable performance while requiring reduced costs and complexity. In hybrid beamforming, digital signal processing is utilized in the baseband to eliminate or mitigate interference, while discrete phase shifters are employed in the RF domain to steer beams.

In general, the problem is to minimize the distance between hybrid and fully digital beamforming for each beam, which is known to be NP-hard. To reduce computations in optimization problems, various methods exist. Various deep learning paradigms, such as the generative adversarial estimation of channel covariance, LSTM in single user scenarios and deep CNNs in the downlink of multi-user settings have been proposed.

And deep supervised learning for beamforming represents a promising, scalable, and statistically robust approach for high mobility scenarios. In these frameworks, the RF signature of the environment and the locations of users/APs are acquired through pilot signals. Additionally, contextual side-information, including user trajectory, past beamforming, situational awareness, and traffic flow, are incorporated during the training phase. Various deep learning paradigms have been put forth, such as the generative adversarial estimation of channel covariance, LSTM in single-user scenarios, and deep CNNs in the downlink of multi-user settings.

Generally, the performance of supervised deep learning algorithms holds promise, yet necessitates extensive labeled datasets for training and is susceptible to unpredictable fluctuations in mm Wave channels, notably caused by prevalent blockage. In addressing this concern, DRL is utilized for hybrid beamforming in point-to-point communications. DRL has its costs as well: its convergence is slow and needs excessive computations, usually provided via cloud computing with high latency and excessive signaling. Besides, stringent time constraints in fast-moving UEs can be met by utilizing edge computing (with significantly less signaling and reduced mobility interruption time) instead of cloud computing, but the challenge is scarceness of computing power at the edge. Therefore, introducing innovative DRL-based strategies with reduced convergence time is crucial for addressing slow convergence.

4.6.2 The Problem Statement of Beamforming with DRL in MIMO

The problem model is based on a Partially Observable Markov Decision Process (POMDP) for beamforming. The problem model aims to address the challenges of beamforming in high-mobility millimeter-wave communication, including signal attenuation and interference.

Specifically, the goal is to maximize EE^{UL} which means the QoS-aware energy efficiency (EE) in the uplink (UL) through beamforming in the presence of interference and signal attenuation. This means achieving high-quality user experience while maximizing energy efficiency.

The problem considers beam steering latency and power consumption. Firstly, we make the system aware of QoS by requiring the data rate of each UE's bandwidth to be higher than its required minimum value. Secondly, we constrain the transmit power per subcarrier, incorporating it into the RF beamforming matrices. Also, we set the Frobenius norm of precoding and beamforming matrices to limit the consumed power.

The problem model aims to maximize the expected reward by adopting an optimal non-stationary policy $\pi^* : \Omega \rightarrow \Delta(\mathcal{A})$. Ω represents the state space, which denotes the set of all possible states the system can be in. $\mathcal{A}$ represents the action space, indicating the set of all possible actions the system can take. $\Delta(\mathcal{A})$ represents the set of probability distributions, which signifies the assortment of probability distributions for potential actions within the specified action space $\mathcal{A}$. The policy π^* specifies a probability distribution over actions to be taken in each possible state. It describes an optimal strategy, where the probability distribution of selecting the best action is determined for each state, aiming to achieve a specific objective or optimization criterion. This is achieved by training a DRL network to learn weight values that set the beam to the desired direction for each instance and location. This enables efficient beamforming in high-mobility millimeter-wave communication, improving user experience and energy efficiency.

4.6.3 POMDP-Based DRL Algorithm Process for Beamforming

A MDP is a mathematical framework used to model decision-making in situations where outcomes are uncertain and involve a sequence of events. MDPs are primarily employed in the field of RL, where an agent learns to make a series of decisions through interaction with an environment to achieve optimal goals.

In such dynamic and fast-changing environments, given that obtaining the total achievable rate for all AP for a given UE is not practical. So, we consider using the partially observed POMDP which is a mathematical framework that models interactions of an agent with an unknown time-varying environment when the agent has limited observations.

We set a virtual central baseband unit as learning agent. Our algorithm uses two clipped Q-functions to train learning agent, one for selecting actions and one for evaluating actions to ensure stability and convergence during training.

The POMDP-based DRL algorithm for beamforming in high-mobility mmWave communications follows a sequential process. It begins with the utilization of a neural network model to learn and infer optimal actions, trained using off-policy DRL. This model considers limited observations to select actions, focusing on precoding and beamforming matrices. The selected action is then executed, applied to the communication system. Subsequently, the agent receives a reward based on system performance, particularly emphasizing QoS-aware EE^{UL}. This reward guides the update of the neural network model's weights and biases through optimization algorithms like gradient descent, aiming to refine the policy for better future decision-making. The iterative nature of this process continues until a stopping condition is met, such as reaching a maximum number of iterations or convergence.

By using the POMDP-based DRL algorithm, the agent can effectively learn and adapt to the dynamic and unpredictable nature of high-mobility mmWave communications. This approach enables the agent to make intelligent decisions for beamforming, maximizing QoS-aware EE^{UL} and improving the overall system performance. And we apply our schemes in two important use cases, named vehicle to infrastructure (V2I) and high speed train (HST) communications, and confirm the feasibility of our algorithm.

In summary, the application of deep learning in beamforming has great potential, and future trends will focus on advanced deep learning models, the application of edge computing, and improvements in DRL algorithms to further enhance system efficiency and performance.

Chapter 5
Learning Traffic and Mobility Prediction

5.1 Graph Neural Networks-Based Network Architecture

In this article, we delve into the complexities and innovative applications of GNNs within network architectures. We offer a comprehensive analysis of GNNs, discussing their characteristics, preferred uses, merits, and drawbacks, imparting a rich understanding of this transformative technology.

Initially, we explore the intricacies of communication networks, systems that link geographically diverse user and terminal equipment, enabling seamless transmission and exchange of information. These systems are established through a range of networking models, including mesh-like, star-shaped, compound, ring, bus, and tree configurations.

Our discussion ventures further as we refine these concepts, assessing the strengths, limitations, and practical illustrations of each network model. By adopting an intuitive and engaging discourse, we strive to make this complex subject matter not only accessible but also interesting for our readers.

Innovatively, we extend the application of GNNs to fortify these network models, underlining our unique contribution in this domain. The crux of our approach hinges on the manipulation of GNN's inherent functionalities–including Graph Convolutional Networks (GCNs), Graph Autoencoders (GAEs), Graph Generative Networks (GGNs), Graph Recurrent Networks (GRNs), and Graph Attention Networks (GATs)–to optimize system performance and enable customized configurations. At its core, our work seeks to define new paradigms in network architecture design, propelled by GNN-based models. We believe that the transformative power of GNNs, combined with a deep understanding of communication networks, can inspire and guide future innovations, redefining network architectures.

H. Zhang and N. Yang, *Deep Learning in Wireless Communications*,
https://doi.org/10.1007/978-981-97-6314-6_5

5.1.1 Network Structure

Computer network is composed of computer system, communication link and network node. It is the field of computer technology and communication technology closely combined, and undertakes two kinds of work: data communication and data processing. From the perspective of logical function, the network can be divided into resource subnet and communication subnet. The resource subnet provides the ability to access the network and process data, and it consists of the main computer system, the terminal controller, and the terminal. The communication subnet provides the network communication function, which is composed of network nodes, communication links and signal transformation equipment. However, the structure of the communication subnet in the network directly affects the network structure. The communication subnetwork can be divided into point-to-point communication channel and broadcast communication channel according to its technology of transmitting data.

Graph convolutional network, referred to as GCN, the simplest GCN has three layers, namely the convolutional layer, the linear layer and the nonlinear activation layer. We mainly have two methods for convolution operation: one is spectral decomposition, that is, spectral decomposition graph convolution (specific); the other is node space transformation, that is, spatial convolution, which takes the spatial characteristics of graph structure data as the starting point to further discuss and explore the representation of neighbor nodes, so that the representation of neighbor nodes of each node is unified and regular. The goal is to make the subsequent convolution calculations easier. When using spatial convolution, we face three key problems: 1. Choosing the right center point. 2. The selection of neighbor nodes is called the size of the sensory domain. 3. How to deal with the characteristics of neighbor nodes, that is, to build an aggregation function that conforms to the characteristics of neighbor nodes.

Graph Generative Network (GNN for short) is a kind of graph generative nervous system used to generate graph data. Its characteristic is that it reorganizes nodes and edges according to certain rules to generate target graphs with specific requirements or properties.

Graph Autoencoder (GAE) is an autoencoder-based GNN that allows semi-supervised or unsupervised learning of graph node information. In the field of deep learning, Auto-encoder (AE) is a class of artificial neural networks that use input information for representation learning.

Graph Recurrent Network (GRN) is one of the earliest GNN models. Compared with other GNN algorithms, GRN usually converts graph data into sequences, which evolve and change recursively during the training process. GRN models generally use Bidirectional RNN (Bi-RNN) and LSTM as network architectures.

The attention mechanism allows a neural network to focus only on the information needed for the task to learn, and it is able to select specific inputs. The introduction of attention mechanism in GNN can make the neural network focus on nodes and edges that are more relevant to the task, improve the effectiveness of training and the ACC of testing, and thus form a Graph Attention Network (GAT).

Combining Graph Neural Networks (GNNs) with DRL can lead to more complex and efficient network architectures, especially for tasks involving relational data such as recommendation systems, social network analysis, image segmentation, and more.

1. Graph-based RL: Graph-based RL combines the capabilities of GNNs with RL techniques to address tasks that involve graphstructured data.

In this paradigm, the environment or problem domain is represented as a graph, where nodes correspond to entities or states, and edges signify relationships or transitions between them. The objective is to train an agent to navigate this graph and make sequential decisions to optimize long-term rewards.

The problem is modeled as a graph, with nodes representing states, objects, or entities, and edges denoting relationships or connections between them. A GNN is employed to comprehend the graph structure and node/edge features. By aggregating information from neighboring nodes and edges, the GNN iteratively updates node representations. A policy network is trained using RL algorithms like Deep Q-Learning or Policy Gradient methods. This network utilizes node representations from the GNN to output actions for the agent. The agent garners rewards based on its actions in the graph environment. The agent's aim is to learn a policy that maximizes cumulative rewards over time. The agent incessantly engages with the graph environment, adjusting its policy network based on received rewards. Simultaneously, the GNN refines its representations as the agent explores the graph.

Graph-based RL is especially beneficial for tasks where entity relationships significantly impact decision-making, such as recommendation systems, social network analysis, and route planning in transportation networks. Through the synergistic fusion of GNNs' graph modeling abilities and RL's decision-making prowess, graph-based RL has demonstrated promise across a range of real-world applications.

2. Graph neural network policies: GNN policies refer to using GNNs as the basis for defining policies in RL tasks. In this setup, GNNs are employed to encode the graph structure and node features, enabling agents to make decisions based on the learned graph representations.

Here is how Graph Neural Network policies work:

Graph Representation: The environment is represented as a graph, where nodes represent entities or states, and edges represent relationships between them. Each node contains features that describe its attributes.

Node Embeddings: The GNN is used to learn node embeddings by aggregating information from neighboring nodes, capturing both node-specific features and relational information in the graph.

Policy Network: The output of the GNN serves as input to a policy network that maps the node embeddings to actions. The policy network can be a neural network that outputs action probabilities or value estimates based on the node representations.

Reward Signals: The agent interacts with the environment based on the actions prescribed by the policy network. Rewards are provided based on the agent's actions, and the goal is to maximize the cumulative reward over time.

Training: The policy network is trained using RL algorithms such as Policy Gradient methods or PG. The GNN is also updated during training to improve the quality of node embeddings.

By utilizing GNN policies in RL, agents can effectively leverage the graph structure and relational information present in the environment. This approach is particularly beneficial in tasks where decisions are influenced by the relationships between entities, such as social network analysis, recommendation systems, and path planning in transportation networks.

Graph Neural Network policies offer a powerful framework for incorporating graph-structured data into RL tasks, enabling agents to make intelligent decisions based on learned graph representations.

3. Graph attention mechanisms: Graph Attention Mechanisms are a crucial component of GNNs that enhance the ability of models to focus on important nodes and edges in a graph structure. Attention mechanisms allow GNNs to assign different weights to the neighbors of each node, enabling the model to selectively aggregate information based on the significance of each neighbor. In a graph, each node has its own feature representation, which captures the attributes or characteristics of the node. Graph Attention Mechanisms use attention weights to determine the importance of each neighbor node when aggregating information for a target node. These attention weights are learned during the training. Attention scores are by measuring the compatibility the features of the node and its neighbors. These scores are then normalized obtain attention coefficients that the importance of each neighbor. The normalized attention are used to aggregate the from the neighbor nodes. Nodes with higher attention weights contribute more to the updated representation of the target node. After aggreg information from neighbors, a learnable update function is applied to compute the new representation of the target node. This updated representation captures both the node's own features and the important information from its neighbors.

Graph Attention Mechanisms have shown significant improvements in various tasks especially in scenarios where relationships between nodes play a crucial role in the decision-making process. By allowin GNNs to focus on relevant information and effectively aggregate features from neighboring nodes, attention mechanisms enhance the model's ability to capture complex dependencies in graph-structured data.

Graph Attention Mechanisms a powerful tool in NNs, enabling models adaptively attend to nodes and edges in graph and improving the performance GNNs in such as node classification link prediction, and graph classification.

4. Combining GNNs with Deep Q-Learning: Combining GNNs with Deep Q-Learning is a powerful approach to address RL tasks in graph-structured data environments. This integration leverages the representational power of GNNs to learn and encode the complex relationships within the graph, while utilizing Deep Q-Learning to make sequential decisions and optimize action selection (Fig. 5.1).

The environment is represented as a graph, where nodes represent states or entities, and edges represent relationships between them. Each node contains features that capture the attributes of the state. A GNN is used to learn node embeddings by

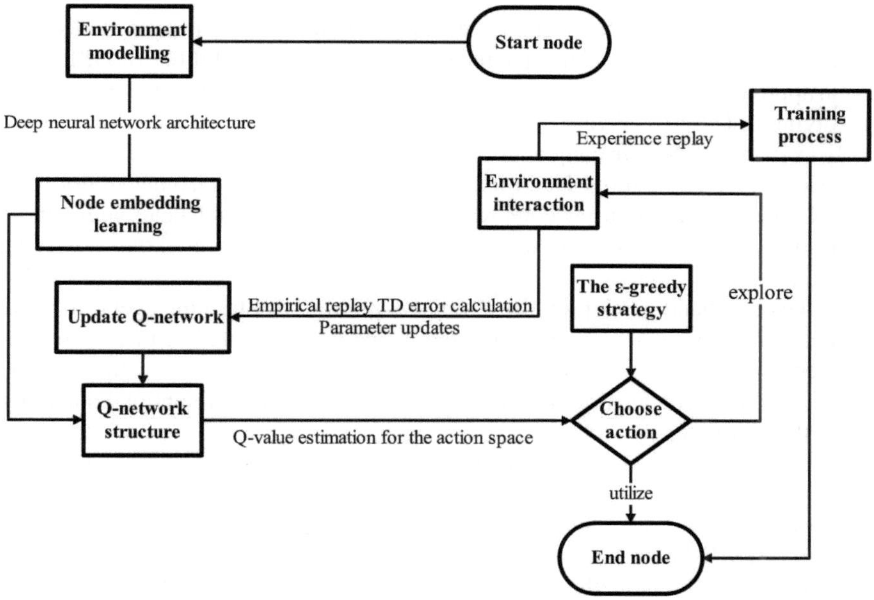

Fig. 5.1 Q-learning in combination with GNNs

propagating information through the graph structure and aggregating features from neighboring nodes. This allows the GNN to capture both node-specific characteristics and relational information in the graph. The output of the GNN serves as input to a Q-network, which is a DNN that estimates the Q-values for each action in a given state. The Q-network takes the node embeddings as input and outputs Q-values for all possible actions. The Q-network is used to select actions based on the estimated Q-values. In Deep Q-Learning, the agent follows an ε-greedy policy to explore the environment and gradually shifts towards a more exploitative strategy as training progresses. To stabilize training and improve sample efficiency, experience replay is often employed. This technique involves storing experiences (state, action, reward, next state) in a replay buffer and sampling mini-batches to train the Q-network. To improve the stability of training, a target Q-network is used to estimate target Q-values during updates. The target network is periodically updated with the weights of the main Q-network. The Q-network is trained using the Bellman equation to minimize the Temporal Difference (TD) error between the predicted Q-values and the target Q-values. The GNN and Q-network parameters are updated through backpropagation to improve the action-selection policy.

By combining GNNs with Deep Q-Learning, agents can effectively learn to navigate graph-structured environments, leveraging the power of graph representation learning and RL. This approach is particularly beneficial in tasks where decisions are influenced by the relationships and dependencies between entities in the graph, such as graph-based pathfinding, recommendation systems, and social network analysis.

The integration of GNNs with DRL enables us to harness the unique of GNNs in modeling graph structures alongside the decision-making capabilities of DRL. This fusion approach facilitates the development of more intelligent and efficient methods for solving tasks across diverse domains.

By leveraging GNNs in conjunction with DRL, we can effectively tackle problems that involve graph-structured data, where entities interact with each other within a network. GNNs excel at capturing complex relationships and dependencies within these graph structures, enabling better understanding and representation of the underlying data.

On the other hand, DRL provides a robust framework for training agents to make sequential decisions that maximize long-term rewards. By combining these two powerful techniques, we can create agents that not only navigate and interpret graph structures effectively but also make informed decisions to achieve optimal outcomes within these structures.

This combined approach has broad applications across various fields, including but not limited to recommendation systems, social network analysis, biological network analysis, and route planning in transportation networks. The synergy between GNNs and DRL holds promising prospects for advancing the state-of-the-art in intelligent task-solving methods and addressing complex real-world challenges.

5.1.2 GNNs for Communication Networks

GNN-based network modeling offers practical advantages that address limitations of previous machine learning-based solutions. These advantages include:

Accurate prediction of network performance: The black-box representation of a generic network model, powered by GNNs, can accurately predict relevant performance metrics at different levels of granularity, such as flow, link, and port statistics. This enables network models to forecast network behavior under various scenarios, including topology changes, upgrades, failures, and new configurations like routing or VNF placement. These predictions are crucial for network control and management tasks, such as what-if analysis and automatic network optimization.

Generalization over graph-structured data: Networks have graph-structured information at multiple levels, making GNNs the most suitable machine learning technique for processing such data. Unlike traditional fully-connected NNs, GNNs are designed to directly capture meaningful patterns from graphs. GNN models leverage a distributed message passing architecture, allowing them to gather local context from graph nodes. This feature enables GNNs to generalize effectively by learning from the experiences of all routers in the network during training. Consequently, the model can apply this learned knowledge to other routers in different networks with varying sizes and structures.

Equivariance to node and edge permutations: GNNs exhibit equivariance to node and edge permutations, meaning theycan identify symmetries or equivalent patterns between network scenarios seen during training and new scenarios encountered during inference. By representing network scenarios as graphs, GNN models can identify clusters within the different types of networks, such as wireless, data centers, and loT, as long as they are represented as graphs. Network topology that are equivalent or similar to those seen during training. This generalization property extends to different types of networks, such as wireless, data centers, and IoT, as long as they are represented as graphs.

The practical advantages of GNNs address fundamental limitations of previous ML-based solutions for network modeling. GNNs enable offline training, allowing models to generalize to new networks. They also facilitate comprehensive testing and certification processes, ensuring the deployability of network models. By leveraging GNNs, data-driven network models can overcome previous limitations and provide accurate predictions for improved network control and management. Offline training is an essential aspect of developing practical data-driven network models. By training the models in a controlled testbed or simulation environment, it becomes possible to expose them to a wide range of scenarios, including extreme cases that are challenging to replicate in real production networks. This allows the models to learn from diverse data and abstract deep insights during the training phase.

The ability of GNNs to generalize to new networks unseen during training is crucial for offline training. GNN models can capture meaningful patterns from graph-structured data, such as network topologies, configurations, and traffic patterns. They can identify symmetries or equivalent patterns between different scenarios, enabling them to make accurate predictions even in unseen networks. This generalization property of GNNs allows the trained models to be readily deployed in customer networks without the need for re-training on the target network.

By leveraging offline training and the generalization capabilities of GNNs, data-driven network models can be developed and deployed more efficiently. This approach saves time and resources by eliminating the need for continuous re-training on specific network instances. Instead, the models can be trained offline on representative datasets, ensuring they are equipped to handle a wide range of network scenarios once deployed.

The deployability of ML-based solutions in critical networking infrastructures requires rigorous testing and certification processes. GNNs offer advantages in this regard by allowing offline training and extensive testing under various operational network scenarios.

During offline training, GNN models can be trained on representative datasets that capture a wide range of network topologies, configurations, and traffic patterns. This enables the models to learn from diverse data and generalize well to unseen networks. After training, the models can be thoroughly tested in controlled testbeds or simulation environments, where their behavior can be evaluated under different operational conditions.

By testing the models offline, it becomes possible to assess their performance, robustness, and limitations across various network sizes and traffic aggregates. This testing process helps generate certifications that define the operational ranges where the model can provide guarantees and reliable performance. These certifications are crucial for ensuring the safe and reliable deployment of ML-based networking products in real-world networks.

Overall, GNNs facilitate the integration of ML-based solutions into the standard commercialization process of networking products. They enable offline training, extensive testing, and the generation of certifications, ensuring that the models are thoroughly evaluated and can be deployed with confidence in critical networking infrastructures.

5.2 Slice Reconfiguration Based on Demand Prediction

Network slicing is a powerful functionality that allows multiple independent networks to coexist on a shared physical network infrastructure. By using different slices of the same spectrum band, organizations can tailor each network slice to meet specific application requirements for security, reliability, and performance. Network slicing relies on technologies such as Software-Defined Networking (SDN), Network Function Virtualization (NFV), and automation to efficiently segment the network and allocate resources. This enables organizations to support different applications, devices, domains, and groups with dedicated network slices. By leveraging network slicing, enterprises can effectively meet Service Level Agreements (SLAs) by ensuring that each application receives the necessary resources and QoS. It offers a cost-effective solution for accommodating diverse application needs on a shared network infrastructure

Network slicing plays a crucial role in the cellular domain by enabling fine-grained control over traffic resources. Each slice has specific resource requirements, QoS parameters, security configurations, and latency requirements. This allows for optimized support for different applications and services. For example, a network slice dedicated to high-definition video streaming would have different characteristics and resource allocations compared to a slice used for monitoring an IoT lighting system. By tailoring network slices to specific use cases, resources are efficiently allocated based on the actual requirements of each application or device. Network slicing helps conserve resources by ensuring that devices receive only the necessary amount of resources they require, preventing overprovisioning where devices have access to more resources than they actually need. By understanding the context and use case of each application, network slicing can allocate resources appropriately, leading to resource optimization. With advancements in core network technologies, such as NFV, network slicing becomes easier to implement, especially in 5G networks. This benefits enterprises, mobile network operators, and managed service providers as they can leverage network slicing to deliver tailored services, improve efficiency, and meet the diverse requirements of different applications and devices.

Network slicing utilizes virtualization technology to create multiple independent networks or slices on a shared network infrastructure. Each slice has distinct characteristics, including latency, throughput, security, and bandwidth. SDN plays a crucial role in enabling network slicing by separating the network control plane from the data plane responsible for handling packets. The control plane defines rules for packet handling on the data plane, allowing the creation of virtual networks or slices with specific characteristics. In the context of 5G networks, network slicing can be seen as an advanced version of Virtual Local Area Networks (VLANs) and extends to both core networks and Radio Access Networks (RANs). Software-Defined RANs (SD-RANs) leverage virtualization to separate and manage traffic on different radio networks, allocate shared resources, and even combine resources from multiple networks when necessary. By implementing network slicing across SD-RANs, network operators can achieve physical separation of traffic, efficient resource allocation, and improved spectrum efficiency. This leads to enhanced resource utilization and enables service providers and private enterprises to deliver more tailored and efficient services compared to previous cellular generations.

Slice reconstruction is a technique used in cloud computing to recover lost or corrupted data. It is an important aspect of data management and ensures the integrity and availability of data stored in the cloud. In cloud computing, data is typically divided into smaller units called slices and distributed across multiple servers or nodes for redundancy and performance reasons. When a slice of data becomes unavailable due to server failure or data corruption, slice reconstruction techniques are used to restore the lost data and ensure its completeness. The process of slice reconstruction involves retrieving the missing or corrupted slices from other nodes or servers in the cloud. This can be done through various algorithms and techniques, such as erasure coding, replication, or recombination. These techniques utilize redundancy and parity information to reconstruct the missing data slices.

Erasure coding is a popular technique used in slice reconstruction, where additional redundant slices are created using mathematical algorithms. These redundant slices contain parity information that can be used to recover lost or corrupted data slices. By distributing these encoded slices across multiple servers or nodes, data reliability and availability are significantly improved.

Replication is another approach used in slice reconstruction, where each data slice is duplicated and stored on multiple nodes. If a slice becomes unavailable, the replicated copies can be used to restore the data. This approach provides high availability but requires more storage space compared to erasure coding. Recombination is a technique that reconstructs missing slices by combining available slices and using coding or algorithmic methods. It involves analyzing the relationships between the existing slices to recover the missing ones. Slice reconstruction plays a crucial role in ensuring the reliability and availability of data in cloud computing. It enables data recovery in case of node failures, data corruption, or other unforeseen events, thus enhancing the overall data management capabilities of cloud systems.

Network slicing is also a technique used in the context of the IoT to enable efficient and customized network services for different IoT applications or use cases. It involves dividing a physical network infrastructure into multiple virtual network slices, each tailored to meet the specific requirements of a particular IoT application. In the IoT, diverse applications with different characteristics coexist, such as smart cities, industrial automation, healthcare, and agriculture. These applications have varying requirements in terms of bandwidth, latency, reliability, security, and scalability. Network slicing allows the network infrastructure to be partitioned to provide dedicated resources and services to each application, ensuring efficient utilization of network resources and optimal performance.

Each network slice is configured with its own set of network functions, including routing, security, and traffic management. It operates as an independent virtual network, providing customized connectivity, SLAs, and QoS to the IoT application it serves. Network slicing enables the isolation and separation of traffic from different applications, ensuring that one application's traffic does not impact the performance or security of another. Network slicing also offers flexibility and scalability by allowing dynamic allocation and reconfiguration of resources based on changing IoT application requirements. It enables organizations to optimize their network resources by allocating them on-demand, reducing costs and improving efficiency. Furthermore, network slicing facilitates the deployment of new IoT services and applications. By providing dedicated network resources, it enables experimentation, innovation, and rapid deployment of new services without affecting existing applications. This helps accelerate the adoption and growth of the IoT ecosystem. network slicing in the IoT enables efficient resource allocation, customization of network services, isolation of traffic, and scalability. It plays a crucial role in supporting the diverse and evolving requirements of IoT applications and optimizing the performance of the IoT infrastructure.

In the context of intelligent cities, network slicing can play a crucial role in providing reliable, efficient, and secure connectivity to support various smart city applications and services. Here are a few examples of how network slicing can benefit intelligent cities:

Enhanced public safety: Network slicing can prioritize critical services such as emergency communications, video surveillance, and real-time analytics for public safety agencies. This ensures that these services receive dedicated network resources, guaranteeing their reliability and low latency.

Efficient traffic management: Intelligent transportation systems rely heavily on connectivity to improve traffic efficiency and safety. Network slicing can allocate specific network resources to traffic management applications, allowing real-time data collection, analysis, and coordination between traffic lights, vehicles, and pedestrians.

Smart grid management: Network slicing can facilitate the integration of renewable energy sources and enable real-time monitoring and control of the power grid. This ensures efficient use of resources and supports smart energy management, demand response, and grid stability.

Enhanced healthcare services: Network slicing can ensure the availability of high-quality, low-latency communication for telemedicine services, remote patient monitoring, and healthcare applications. This enables faster access to healthcare resources and improves patient care in intelligent cities.

Innovative IoT Applications: Intelligent cities rely heavily on interconnected devices and applications. Network slicing can allocate dedicated resources for specific IoT use cases, such as smart buildings, environmental monitoring, waste management, and parking systems, ensuring reliable connectivity and efficient data transmission. By leveraging network slicing, intelligent cities can optimize their network infrastructure to cater to diverse applications and services, ensuring efficient resource utilization, low latency, and enhanced user experience. However, it's crucial to ensure proper security measures and coordination between different network slices to prevent interference and maintain overall network performance.

Deep learning can play a significant role in enhancing network slicing Network slicing is a technique used in 5G networks to divide a physical network infrastructure into multiple virtual networks, each tailored to requirements. Deep learning algorithms be utilized in network slicing to optimize resource allocation, enhance efficiency, and improve overall network performance. Deep learning models learn from vast amounts of data, enabling them to understand complex patterns and make accurate predictions. By analyzing network traffic, user behavior, and other relevant data, deep learning algorithms can identify traffic patterns, predict future demand, and allocate network resources accordingly. This dynamic resource allocation allows network operators to maximize network capacity, reduce latency, and ensure a seamless experience. It can be applied to network within network slicing. By analyzing network traffic and user behavior deep learning algorithms can detect and prevent cybersecurity threats in real-time. This in enhancing the security and integrity of the network slices, protecting sensitive data and ensuring uninterrupted service delivery. Specific aspects are as follows:

1. Traffic prediction: Deep learning algorithms can be utilized to analyze historical data and predict traffic patterns in IoT networks. By understanding the expected traffic load for different IoT applications, network slicing can be optimized to allocate appropriate resources to each slice accordingly.

2. QoS optimization: Deep learning models can be trained to analyze real-time data from IoT devices and predict the QoS requirements of various applications. This information can then be used to dynamically allocate network resources to ensure that each slice meets its specific QoS objectives.

3. Anomaly detection: Deep learning algorithms can be used to identify abnormal behavior or unexpected events in IoT networks. By monitoring network traffic device data, deep learning models can detect anomalies that may impact network slicing performance. This allows for timely and optimizations to maintain the desired network performance.

4. Security and privacy: Deep learning techniques can be applied to develop intrusion detection systems (IDS) IoT networks. These models can learn patterns of malicious activities and help protect different slices from threats. Deep learning can also be leveraged to enhance privacy by developing models that detect and prevent unauthorized access or breaches within network slices.

5. Resource optimization: Deep learning can assist in optimizing the allocation and utilization of network resources in IoT network slicing. Through analysis of historical data and real-time monitoring, deep learning models can learn to allocate

resources efficiently based on the specific needs of different slices, allowing for better resource management and improved overall performance.

Overall, deep learning empowers network slicing with intelligent and adaptive capabilities, enabling network operators to optimize resource allocation, enhance network efficiency, and improve network security.

5.3 Learning Cellular Traffic Prediction

The cellular network, a vital communication network, offers call, message, and data services to end users within the coverage of BS. Throughout its history, the cellular network has seen remarkable evolution and progress, continuously enhancing mobile communication services and data transmission rates. In the late 1970s, the first-generation cellular network (1G) emerged, providing analog-based voice communication at a considerable cost. However, it suffered from limitations in network coverage and mobile phone battery power, resulting in subpar service quality, including frequent call drops. Subsequently, the analog transmission system underwent an upgrade to a digital transmission system, known as 2G, in the 1990s. This upgrade significantly improved both the reliability and security of the service. Additionally, the global system for mobile communication of 2G introduced the convenient short message service (SMS). The 2G system implemented both time division multiple access and code division multiple access (CDMA) technologies. During the transition from 2G to 3G, a 2.5G network utilizing general packet radio service (GPRS) facilitated internet communication. The 3G network, powered by technologies like the universal mobile telecommunication system and CDMA2020, offered enhanced mobile internet connectivity, enabling various types of services such as web browsing, email, image, and video transmission. Compared to its predecessor, the 3G network, the 4G networks, such as worldwide interoperability for microwave access and long-term evolution (LTE), showcased significant speed improvements. The 4G network also enabled mobile broadband transmission services, including high-quality audio and video streaming.

As the nascent generation of cellular networks, currently in its early stages of commercial deployment, the 5G network encompasses three distinct application scenarios: enhanced mobile broadband (eMBB), massive machine type communications (mMTC), ultra-reliable and low latency communications (URLCC). While data transmission rate remains a crucial factor, it is not the sole criterion. The specific goals of each scenario, such as low battery consumption and improved connectivity, are equally significant. To accomplish these objectives, a combination of diverse communication technologies and AI technologies is employed. AI plays a pivotal role in optimizing the 5G network, allocating resources optimally, unifying acceleration of the 5G physical layer, and facilitating end-to-end joint optimization of the physical layer.

Within the domain of cellular prediction, a classification scheme has been introduced to delineate different workflows and models. This classification highlights four distinct workflows: direct-prediction, classification-then-prediction, decomposition-then-prediction, and clustering-then-prediction. Each of these workflows utilizes specific data preprocessing techniques to achieve accurate predictions. In terms of models, they can be broadly categorized into three types: statistical models, machine learning models, and deep learning models. It is worth noting that deep learning models, which epitomize the forefront of AI, have emerged as the leading solutions in this field, showcasing their prominence and efficacy.

We begin by classifying cellular traffic problems into two main types: temporal prediction problem and spatiotemporal prediction problem, arising from distinct scenarios illustrates the temporal prediction problem, where a single base station is considered, and only the traffic generated by users or devices connected to this specific base station is taken into account. In this straightforward scenario, the prediction relies solely on the temporal dependencies within the historical traffic data. On the other hand, the spatiotemporal prediction problem, which involves users moving and transitioning between different BS through handover processes. This more complex problem considers the traffic across multiple BS or regions, incorporating both spatial and temporal dependencies. In certain specialized instances of the spatiotemporal prediction problem, the objective may be to predict the entire traffic distribution within a given area, or solely focus on hotspot areas.

Both the temporal prediction problem and spatiotemporal prediction problem can be framed as supervised learning problems through the utilization of moving windows to generate different input and output pairs. The traffic data collected is represented as a univariate time series, and the prediction of future time step values is based on historical data from past time steps, with a fixed length. The moving windows, serve the purpose of generating both the input historical data and the prediction targets.

In real-world cellular networks, various metrics can be employed to measure traffic intensity. These metrics include, but are not limited to, SMS, call service, internet usage service, physical resource block (PRB) utilization data, and the number of connected users. When only one metric or the aggregated traffic value is collected and utilized, it simplifies into a univariate prediction problem. However, by incorporating multiple metrics and simultaneously predicting them, it becomes a more complex multivariate prediction problem. In such cases, it is essential to consider the dependencies between different services, applications, or users when measuring the traffic volume for distinct services or applications. This consideration of service-wise or application-wise dependencies adds an additional layer of complexity to the prediction problem.

In all cases, the timeline is divided into periodic time slots with varying levels of time granularity. Most studies utilize a time period ranging from five minutes to one hour, especially when utilizing open datasets. While rare, some studies employ extreme values, such as millisecond resolution for predicting PRB utilization data. The focus here is on the single-step prediction problem, aiming to predict traffic for the next time slot. However, the formulations presented can be easily extended to the multi-step prediction case, where the prediction target includes the traffic volume for multiple future time slots.

In practice, various data preprocessing steps are needed before the core prediction task. Now, let's look at a few common methods of data preprocessing. Four types of general prediction workflows are considered–direct prediction, post-classification prediction, post-decomposition prediction, and post-clustering prediction–which require the use of different data preprocessing techniques.

(1) Direct prediction

In the context of direct prediction, historical data and forecast targets are typically formatted as time series or input vectors. To prepare the data for analysis, some general data preprocessing techniques are often applied, such as data scaling through standardization or min-maxl normalization. These techniques help to ensure that the data is on a similar scale and can be effectively compared. However, in real-world scenarios, data collection processes may not always be perfect, leading to missing data. In such cases, data filling techniques are necessary to address the missing observations. There are several options available for data filling, depending on the complexityl of the problem. For simple cases, forward fill or moving average methods can be used to fill in missing values. Forward fill involves carrying forward the lastl observed value, while moving average replaces missing values with the average of neighboring values. In more complex scenarios, techniques like Bayesian Gaussian tensor decomposition can be considered. This method uses a probabilisticl approach to estimate missing observations based on the observed data and their relationships. Normalization, on the other hand, is a technique used to transform data to a common scale. It does not change the underlying pattern of the data but adjusts the range of values. This is often done to ensure that different variables or time series can be compared.

(2) Post-classification prediction

In the post-classification prediction workflow, the traffic data for different applications or services is collected from the same data source, such as raw data packets from a base station or user device. Deep packet inspection technology is utilized to extract the specific details of the transmitted data, which serves as the foundation for traffic classification. Machine learning and deep learning models are then employed to classify the data packets into specific applications or services, such as email, SMS, video streaming, audio chat, or video calls. Once the traffic data is classified, it is aggregated separately for each application or service. Multiple predictive models are then constructed to forecast future traffic for individual applications. This means that distinct models are created for each application or service, taking into account their unique traffic patterns and characteristics. There are two main benefits of classifying traffic before forecasting. Firstly, the internal pattern of traffic for a single application tends to be more consistent and evident during subsequent forecasting compared to the combined and aggregated total traffic. This makes it easier for the predictive model to achieve better performance by focusing on the specific patterns of each application. Secondly, by extracting detailed observations of data usage from different applications, measurements can be designed accordingly. For instance, when additional transmission bandwidth is required for more important applications, the quality of video streams can be reduced. Overall, the post-classification prediction workflow enhances forecasting ACC by considering the specific patterns and characteristics of individual applications or services. It allows for more targeted and accurate predictions, taking into account the unique requirements and behavior of each application.

(3) Post-decomposition prediction

In the post-decomposition prediction workflow, a single input traffic time series is decomposed into multiple components. Each component represents a different aspect or pattern within the overall traffic data. These components are then used as inputs for separate predictive models, where each model is responsible for predicting a specific component. Finally, the predictions from all the individual models are combined to obtain the final prediction. Unlike traffic classifications, where components correspond to specific applications or services, the components in the decomposition workflow do not have a physical meaning. They are extracted to capture different patterns or characteristics present in the traffic data. By decomposing the time series into these components, it becomes easier to model and predict each component individually, potentially leading to better prediction ACC. The decomposition process can be performed using various techniques, such as time series decomposition methods like seasonal decomposition of time series (STL) or empirical mode decomposition

(4) Post-clustering prediction

In the post-clustering prediction workflow, the input traffic time series data is collected from different sources, such as different BS or cells. The purpose of clustering is to group these sequences based on their similarity. By doing so, a small number of predictive models can be built instead of creating a separate model for each base station, which can be computationally expensive, especially in large spatial regions with a high number of BS. Clustering helps in improving prediction performance by grouping similar time series data together. The more similar time series are in the same cluster, the more data the corresponding prediction model will have as input. This increased amount of data can enhance the prediction performance and prevent overfitting issues. Clustering ensures that the training data within the same cluster is consistent, which can lead to better predictions. Overall, the post-clustering prediction workflow is beneficial for managing the computational burden of building individual models for each base station and improving prediction ACC by leveraging the similarities within clusters.

5.4 Traffic and Mobility on the Internet of Vehicle

The IoV encompasses the integration of vehicles with the internet and other vehicles to enable communication and data sharing among vehicles. The IoV is revolutionizing the transportation industry by enabling vehicles to communicate with each other and with the surrounding infrastructure. Its significance lies in improving road safety through real-time information exchange, enhancing traffic efficiency with intelligent routing and adaptive control, reducing carbon emissions by optimizing travel patterns, and offering a more connected experience with advanced assistance systems and personalized services. Io also plays a key role in smart city integration promoting seamless mobility and resource utilization while driving economic benefits cost savings and improved efficiency. Ultimately, IoV shaping the future of transportation towards safer, more efficient, and sustainable mobility solutions.

A crucial aspect of IoV is traffic and mobility management, which utilizes deep learning techniques to analyze and exchange real-time information between vehicles and infrastructure. Deep learning algorithms have the capability to process and analyze large volumes of data collected from sensors, cameras, and GPS devices to understand traffic conditions, forecast congestion, and optimize routes. By disseminating this information to other vehicles and infrastructure, drivers can make informed decisions that enhance traffic flow, alleviate congestion, and improve safety.

By leveraging deep learning techniques and the capabilities of connected vehicles, traffic and mobility on IoV can be greatly enhanced. Deep learning algorithms can analyze real-time and historical data from vehicles, infrastructure, and other sources to optimize traffic flow and improve transportation efficiency.

Traffic prediction in the context of IoV can greatly benefit from deep learning techniques. Deep learning algorithms are good at analyzing large amounts of data to identify complex patterns and make accurate predictions. By leveraging the connectivity and real-time data exchange capabilities of IoV, deep learning models can provide valuable insights and predictions about traffic conditions. To predict traffic conditions using deep learning

Fusing two deep learning application scenarios, traffic flow prediction and intelligent Iot anomaly detection and prediction, can build a more intelligent and integrated system. Here's a possible fusion:

Data acquisition and preprocessing: Collect real-time and historical data from various data sources such as vehicles, traffic sensors, IoT devices, weather stations, social media platforms, etc. This data will include factors such as traffic flow, vehicle movement, road conditions, weather information, sensor data, etc. This data is cleaned, denoised, handled for missing values, and normalised and feature engineered in preparation for model training and prediction.

Model selection and training: Select a suitable deep learning model such as RNN or CNN for traffic flow prediction and anomaly detection and prediction. RNN can be used to capture the temporal dependence of time series data and CNN can be used to process image data. The model will learn traffic patterns and abnormal behavior patterns at the same time to improve the intelligence and flexibility of the system.

Model evaluation and optimization: The test dataset is used to evaluate the trained deep learning model and measure its performance in traffic flow prediction and anomaly detection and prediction. According to the evaluation results, the model is optimized and adjusted to improve its ACC and stability.

Prediction generation and response: Once the model is trained, it can be leveraged to generate multiple predictions including traffic flow predictions and anomaly predictions. The system can make real-time traffic management and safety response according to these prediction results, and improve the efficiency and safety of urban traffic system.

By integrating traffic flow prediction and intelligent Iot anomaly detection and prediction, urban traffic systems can be monitored and managed more comprehensively, potential problems can be prevented in advance and emergencies can be quickly responded to, which brings more possibilities for urban traffic management and intelligent iot applications. The ACC reliability of traffic prediction using deep learning depends on the quality and quantity of the collected data, the complexity of the modalities, and the training process of the selected deep learning model. Continuous and periodic retraining of the model ensures that its performance remains optimal as traffic conditions evolve. By accurately predicting traffic patterns, deep learning models can help inform traffic management strategies, optimize the routes of individual vehicles or fleets, and assist transportation authorities and drivers alike in making decisions.

In IoV, anomaly detection and prediction play a vital role in ensuring the safety and efficiency of traffic and mobility. Deep learning technology can also be used to effectively identify and predict anomalies in the IoV ecosystem. Continuous evaluation and retraining of deep learning models with the latest data ensures that their performance remains reliable when new anomalies appear or traffic conditions change. Moreover, integrating real-time data from connected vehicles and infrastructure in IoV enables more accurate and proactive anomaly detection and prediction systems. By utilizing deep learning for anomaly detection and prediction in IoV, traffic management systems can quickly identify abnormal situations, such as accidents, congestion, or faults. This enables fast incident response, traffic diversion, and enhanced safety measures, ultimately helping to improve and make the traffic and travel experience more efficient.

In the IoV ecosystem, intelligent routing and navigation systems play a crucial role in optimizing traffic and mobility. Deep learning techniques can be applied to develop complex models to improve the ACC and efficiency of routing and navigation. Here's how deep learning can be used for intelligent routing and navigation systems in IoV. Data is generated by collecting comprehensive data from various sources, including historical traffic patterns, real-time sensor data, road conditions, weather information, as well as GPS traces. These data form the basis for training deep learning models. The collected data were cleaned and preprocessed to remove noise, handle missing values, and normalize the data. In addition, feature engineering techniques can be applied to relevant information such as road network topology, traffic signal timing, or historical traffic congestion levels. The appropriate deep learning model is then selected for intelligent routing and navigation. CNN can be used to process spatial data, such as maps of road networks or images from traffic cameras. RNN can capture temporal dependencies, which are important for analyzing sequential data such as GPS trajectories or historical traffic patterns. The preprocessed data is then used to train the selected deep learning model. This involves input data such as GPS coordinates, start and end points, and historical traffic information, and training a model to predict the best route and estimated travel time. Reinforcement techniques can also be employed to optimize the decision-making process of the model based on rewards and feedback from simulated or real-world environments. Finally, the trained deep learning model is validated using evaluation metrics such as route ACC, travel time prediction ACC, or comparison with ground truth data. Cross-validation or deploying a simulation environment can help evaluate the performance of a model and identify areas for improvement.

Once the model is trained and validated, it can be used in real-time routing and navigation systems. The deep learning model takes into account the current traffic conditions, road constraints, and user preferences to provide optimal route suggestions, dynamic re-routing in response to traffic congestion, and accurate travel time estimation. The model can also incorporate real-time data from connected vehicles and infrastructure to further enhance its recommendations.

As traffic patterns and road conditions change over time, it is important to continuously update and retrain deep learning models. This will ensure that the smart route and navigation system remains up-to-date and provides accurate guidance. By leveraging deep learning for intelligent routes and navigation in IoV, drivers can experience higher travel efficiency, reduced congestion, and safer navigation. In addition, transportation authorities can use these systems to optimize traffic flow, reduce environmental impact, and enhance the overall travel experience in the IoV ecosystem.

In the IoV environment, data privacy and network security are important considerations to ensure the protection of sensitive information and prevent unauthorized access. Deep learning technologies can contribute to enhancing data privacy and network security in the following ways:

Secure data transfer: Deep learning models can be trained to encrypt and securely transfer data between vehicles and infrastructure. Techniques such as Secure Socket Layer (SSL) encryption, secure multiparty computation, or homomorphic encryption can be applied to protect data in transmission against unauthorized access or interception.

Anomaly detection: Deep learning models can analyze patterns in network traffic and identify anomalies that may indicate malicious activity or cyber threats. By training on known normal behavior patterns, these models can detect suspicious or abnormal network behavior and issue alerts to help identify potential network security threats.

Intrusion detection and prevention: Deep learning models can be deployed to monitor and analyze network traffic in the IoV ecosystem. By continuously monitoring incoming and outgoing packets, these models can detect patterns associated with known network attacks or suspicious activity, enabling rapid response and preventing unauthorized access or data disclosure.

Security authentication and access control: Deep learning models facilitate security authentication and access control mechanisms in IoV. By leveraging techniques such as biometrics or behavior analysis, these models can authenticate and authorize legitimate users or vehicles, preventing unauthorized access and identity theft.

Privacy-preserving data analytics: Deep learning techniques such as federated learning or differential privacy can be used to train models while preserving the privacy of sensitive data. These approaches allow data analytics to be performed locally on a single vehicle or infrastructure node without transferring raw data, preserving user privacy while still enabling valuable insights and collaborative learning.

Threat intelligence and response: Deep learning models can be trained on a large corpus of cybersecurity threat intelligence data. These models can identify the patterns, characteristics, and metrics of known cyber threats, so as to more quickly identify and respond to cyber attacks or vulnerabilities emerging in the IoV ecosystem.

Continuous model monitoring and updating: Deep learning models for cybersecurity can be continuously monitored and updated to adapt to changing threats and vulnerabilities. Regular updates ensure that the model remains effective in detecting the latest cyber attacks and can effectively mitigate emerging risks.

It is important to note that while deep learning can facilitate data privacy and cybersecurity in IoV, a holistic approach is necessary. This includes strong encryption practices, secure network infrastructure, regular security audits, user awareness training, and a regulatory framework to protect user privacy and ensure responsible data handling in the IoV ecosystem.

In summary, combining deep learning techniques with IoV can revolutionize transportation and mobility. By anticipating traffic patterns, detecting anomalies, optimizing routes, and facilitating intermodal transportation, deep learning can contribute to safer, more efficient, and sustainable transportation systems.

Chapter 6
Software Defined Networking

6.1 A Balance Between Flexibility and Manageability

Software Defined Networking (SDN) is a new network architecture and network management method. In a traditional network, network devices (such as switches and routers) are responsible for data transmission and control plane processing. Network administrators need to configure these devices one by one, which makes the control and management of the entire network complicated and rigid. SDN separates the control plane and data plane of the network so that the network control logic can be abstracted from the physical device and centralized in a centralized software called controller for management and configuration.

SDN essentially solves the problem of balancing the flexibility and manageability of the network. In traditional networks, there is a trade-off between flexibility and manageability, which is either sufficiently flexible but less manageable, or manageable but less flexible. The traditional network architecture is usually fixed and decentralized without a centralized network controller, so although the management of the network is relatively simple, it is not flexible when the network structure needs to be adjusted or new requirements need to be introduced. To ensure flexibility, you have to increase the complexity of management.

SDN achieves a balance between flexibility and manageability by separating the network control plane and the data plane into a centralized network controller.

The algorithms to achieve this balance include the shortest path algorithm, the minimum bandwidth path algorithm and the minimum traffic scheduling algorithm. The goal of the shortest path algorithm is to find the shortest path between two nodes so that the cost required to pass through the path is minimal, so as to optimize resource allocation in the communication network, minimize latency or maximize bandwidth. The principle of the minimum bandwidth path algorithm is similar to that of the minimum traffic scheduling algorithm. Formulas and pseudo-code examples are presented below to illustrate. (Take shortest path algorithm as an example.)

H. Zhang and N. Yang, *Deep Learning in Wireless Communications*,
https://doi.org/10.1007/978-981-97-6314-6_6

Dijkstra's algorithm is a greedy algorithm for finding the shortest path from the source node to all other nodes in a graph. The main idea of the algorithm is to start from the source node and gradually expand to the nodes where the shortest path has not been determined until the shortest path of all nodes is determined.

The formula $dist[v] = min(dist[v], dist[u] + weight(u, v))$ describes the key steps in Dijkstra's algorithm, where: $dist[v]$ represents the current shortest distance from the source node to node v. $dist[u] + weight(u, v)$ represents the path length from the source node through node u to node v, that is, the distance from node u to node v plus the current shortest distance from the source node to node u. The $min()$ function is used to compare the smaller values between the two to update the shortest distance of the node v.

The mechanism of the shortest path algorithm is to initialize the shortest distance of the source node to 0, and the shortest distance of other nodes to infinity. Then, the node with the shortest distance is selected from the nodes whose shortest path has not yet been determined as the current node. Then, all adjacent nodes of the current node are relaxed. If the shortest distance is shorter than the currently known shortest distance, update the shortest distance and repeat the steps until the shortest distance is determined for all nodes.

In software-defined networks, the shortest path algorithm will calculate the shortest distance between the source node and the target node, and these paths can be based on different indicators, such as the shortest distance, minimum delay or maximum bandwidth, so as to achieve the separation of the data plane and the control plane required by the software defined network.

A software-defined network can schedule traffic based on different indicators. For example, the controller calculates the shortest path and sends it to the data plane device through the southbound interface between the control plane and the data plane. Once the path information is sent to the data plane device, the data packet is routed according to the path information. The upstream entries on the device on the data plane are configured to match and forward traffic based on the characteristics of the received traffic object. At the same time, in order to ensure the flexibility of the network, when the traffic mode changes, the controller can re-run the shortest path algorithm to update the path information, thus realizing the global view and control of the network, while the data plane device is only responsible for the actual data forwarding. This separation ensures the flexibility, programmability and management efficiency of the network. Therefore, the flexibility and manageability of the network are balanced.

In traditional networks, traffic routing is usually handled by routing protocols on routers and switches, which can be distance vector protocols (such as RIP) or link state protocols (such as Open Shortest Path First(OSPF)). These protocols select routing paths based on network topology and predefined policies.

RIP uses distance-based routing, where distance is a metric, usually the number of hops. Each router maintains a routing table that contains information about the distance to the destination network. Routers update these routing tables by periodically exchanging routing update messages. OSPF uses the link state protocol, and each router maintains a topology database that contains the topology information of the entire network. Link state update messages are periodically exchanged between routers to build a topological database of the entire network and calculate the shortest path using Dijkstra's algorithm.

In fact, both traditional networks and software-defined networks use the shortest distance algorithm for calculation, but the centralized control of software-defined networks determines that it will have more advantages. In traditional network routing protocols, each router only knows the status of its immediate neighbors without a global view of the entire network. At the same time, their route calculation is usually based on distributed algorithms, different from software-defined networks, such as distance vector protocol and link state protocol, which need to be carried out on each router rather than centralized control, and for large networks, the complexity of route calculation may be high, resulting in a long time and slow update.

In contrast, software-defined networks have a centralized controller that can obtain a global view of the entire network, and use the shortest path algorithm for routing calculation at this time, which can adapt to network changes more quickly, achieve rapid network convergence and traffic adjustment. These centralized management and control make software-defined networks stand out in terms of the balance between manageability and flexibility.

SDN solves two key technical bottlenecks in traditional networks, that is, it makes the network programmable and ensures the balance between centralized control and distributed data plane. Traditional network devices are often statically configured, which is difficult to flexibly adjust and manage according to requirements. One of the keys of SDN technology is network programmability, that is, by separating the network control plane from the data plane, network administrators can use standardized programming languages (such as OpenFlow etc.) to dynamically control network traffic. This enables more flexible and intelligent network management to meet the requirements of different applications and services. In this case, the architecture of the separation of control plane and data plane in SDN makes network management more flexible, but also brings the centralization of control plane and the distribution of data plane. This architecture can improve the manageability and flexibility of the network in some ways, but it can also introduce single points of failure and performance bottlenecks. Therefore, it is necessary to find a balance point between centralized control and distributed data planes to ensure network stability, reliability and performance. In this process, intelligent technology has played a key role, and the specific effects of intelligent technology represented by machine learning will be introduced in detail next.

6.2 The Role of Intelligent Technology Represented by Machine Learning

SDN uses intelligent technology to solve the problem of network programmability, mainly through the following five ways: First, intelligent controller. Second, intelligent routing and traffic engineering. Third, intelligent policy management. Fourth, intelligent network monitoring and optimization. Fifth, intelligent applications and services.

The controller in SDN architecture is the intelligent core of the whole network. The controller can use intelligent algorithms to analyze network traffic, device status, application requirements and other information, so as to realize intelligent programming of network behavior. For example, the controller can dynamically adjust network traffic routing and optimize load balancing based on real-time traffic. On a traditional network, the communication logic between network devices is hard-coded inside the devices. Therefore, modifying the configuration requires device-by-device operations, which is difficult and time-consuming. Through the characteristics of centralized control and intelligent flexibility, it can adapt to network changes and requirements more quickly, reducing the cost of manual configuration and management. The classic OpenDaylight Controller and Open Network Operating System (ONOS) integrate algorithms and applications through intelligent technology to take advantage of this principle. In short, OpenDaylight is an open source SDN controller platform, it provides a rich API and plug-in mechanism, can support a variety of intelligent algorithms and application integration, so as to achieve intelligent programming and management of the network. ONOS is another open source SDN controller platform, which adopts distributed architecture, supports the collaborative work of multiple controllers, and provides a wealth of intelligent algorithms and applications, which can realize intelligent control and optimization of the network.

The centralized controller in SDN can realize intelligent routing by monitoring the topology and traffic of the whole network and using intelligent algorithms. The controller can dynamically adjust the routing policy according to the current network status and traffic load, and select the optimal path to transmit data packets. This prevents network congestion and improves network performance and reliability. In addition, a centralized controller in SDN can engineer the traffic in the network according to the policies and goals of the network administrator. For example, the controller can dynamically adjust the traffic path and priority according to the traffic load and QoS requirements to optimize the utilization efficiency and QoS of network resources. In this way, the intelligent control and management of network traffic can be realized, and the performance and efficiency of the network can be improved. Dijkstra algorithm plays an important role in SDN and is one of the bases of intelligent routing. Dijkstra algorithm is a classical algorithm used to calculate the single source shortest path in weighted graphs. In SDN, when it is necessary to calculate the optimal packet transmission path according to the topology and traffic situation of the network, Dijkstra's algorithm can help achieve this goal. In SDN, Dijkstra algorithm is used to calculate the single source shortest path in the network topology, so as to achieve the optimal packet transmission path. The SDN controller obtains network topology information and constructs network topology diagram. Then, Dijkstra algorithm is used to calculate the shortest path and consider the weight index on the path. Finally, the routing table is updated and the optimal path information is sent to the device to realize intelligent routing. This process enables SDN to dynamically calculate the optimal path, meet the requirements of network performance and traffic load, and improve network performance and efficiency.

SDN use access control lists (ACLs) to implement intelligent technical management to improve network security and management efficiency. The SDN controller dynamically generates ACL rules based on real-time network status and traffic and delivers them to network devices. For example, if a host generates an abnormally large amount of traffic, the controller automatically delivers ACL rules to restrict the host's network access to avoid adverse impact on the network. This flexible ACL delivery mechanism enables the network to quickly respond to various security threats, enhancing network security and management efficiency. In addition, the SDN controller can manage ACL rules according to the policies and service requirements set by the administrator. The administrator can define ACL rules through the user interface or API provided by the controller, and set parameters such as the priority and application scope of the rules. The controller automatically generates ACL rules based on these policies and sends them to network devices. This policy-based ACL management method helps administrators control network access flexibly, improving network security and management efficiency. In addition, the SDN controller can use ACL rules to classify and prioritize traffic. Administrators can set ACL rules to identify specific traffic (such as video and voice traffic) and specify priorities for these traffic. The controller classifies and marks traffic based on ACL rules and delivers the traffic to network devices for intelligent management and control. This traffic classification and priority control can help the network realize intelligent scheduling when dealing with various traffic types and improve network performance and service quality.

Real-time flow control through SDN and intelligent technology is also an important means to improve network performance and service quality. One of the most classic mechanisms is to use SDN controllers to collect network traffic information and then use intelligent technologies such as deep learning or rule-based classification algorithms to identify and classify traffic and distinguish real-time traffic from other types of traffic. Then, the identified real-time traffic is marked preferentially, usually using a QoS marking mechanism, such as Differentiated Services (DiffServ) or 802.1p, so that network devices can process traffic based on the priority. Secondly, based on traffic identification and priority marking, the SDN controller can dynamically allocate bandwidth to ensure that real-time traffic gets enough bandwidth to meet its delay requirements, so as to achieve dynamic bandwidth allocation. For example, TCP congestion control algorithms (such as TCP Tahoe, TCP Reno, TCP NewReno, TCP Vegas, etc.) are technologies that realize adaptive traffic control by monitoring network congestion and adjusting the sending rate. And another very classic example is fault recovery, SDN controller can use intelligent technology to monitor the fault in the network, and take corresponding measures to restore real-time traffic transmission. It is worth noting that SDN itself can realize the programmability of the network and its branch function through a special architecture, in which intelligent technology plays an auxiliary role.

In the combination of SDN and intelligent technology, one of the most important aspects of network optimization and security strategy is intelligent security analysis. Through intelligent security analytics, SDN controllers can monitor network traffic and security events in real-time, analyze network behavior patterns using intelligent algorithms and machine learning techniques, and detect abnormal traffic and potential threats. This kind of intelligent security analysis can effectively identify and block the malicious behavior in the network, including intrusion attacks, malicious software spread and other security threats. Intelligent security analysis can not only detect security incidents in time, but also provide more intelligent security response and defense mechanisms. Based on the intelligent security analysis results, the SDN controller dynamically adjusts network security policies and takes targeted defense measures to prevent malicious traffic from entering the network and protect critical services and sensitive data.

In SDN, another difference from the traditional network architecture is that it adopts a centralized control mode. In traditional networks, network devices (such as routers and switches) are responsible for data forwarding and processing, and the decision-making process is decentralized among each network device. In SDN, the control logic of the network is centralized into a centralized controller, which manages the behavior of the entire network by communicating with network devices. This enables faster deployment and centralized management of SDN. However, centralized control also faces some challenges, including single points of failure, performance bottlenecks, and security issues. Therefore, in actual deployment, it is necessary to consider how to address these challenges to ensure the reliability and security of centralized control. Next, it will explain how SDN realizes centralized control through intelligent technology from four aspects: distributed control plane, intelligent routing and load balancing, distributed data plane collaboration and intelligent network monitoring and management.

Centralized control uses a single controller to manage the entire network, which has the advantages of centralized management, flexibility and programmability, but it also has the problem of single point of failure and performance bottleneck. The distributed control plane adopts a distributed management network with multiple controller nodes, realizes state consistency and decision coordination through RAFT protocol and Paxos algorithm, overcomes the shortcomings of centralized control, and improves the reliability and performance of the network. RAFT protocol and Paxos algorithm are two common consistency protocols used to achieve state consistency in distributed systems. They play an important role in the distributed control plane, ensuring state synchronization and decision consistency among multiple controller nodes, so as to achieve unified management and control of the network. In SDN, consistency protocols such as RAFT protocol and Paxos algorithm are widely used in the design and implementation of distributed control plane to achieve state synchronization and decision coordination among multiple controller nodes. These protocols ensure state consistency between controller nodes, so as to achieve unified management and control of the entire network. With the support of consistency protocols such as RAFT protocol and Paxos algorithm, SDN can better realize distributed control plane and improve network reliability and performance.

In SDN, based on the global network topology and traffic information, the centralized controller can use intelligent routing algorithms to calculate the optimal path, the shortest path or the best path from the source node to the destination node. The centralized controller also implements load balancing of network traffic, that is, traffic is distributed over multiple equal paths to achieve balanced utilization of network resources and performance optimization. Intelligent routing and load balancing are usually implemented by ECMP (Equivalent multipath) and shortest path algorithm. ECMP is a hash function based load balancing algorithm that distributes traffic over multiple equivalent paths to achieve load balancing. Specifically, ECMP achieves load balancing by taking information such as the packet's source IP address, destination IP address, and port number as input, using a hash function to calculate a value, and then sending the packet to the path corresponding to that value. The shortest path algorithm is an algorithm used to calculate the shortest path between two nodes in a network, including Dijkstra's algorithm and Bellman-Ford's algorithm. Based on the greedy strategy, Dijkstra's algorithm starts from the initial node, gradually expands to other nodes in the network, and continuously selects nodes on the shortest path to join the set of shortest paths until the target node is reached. Bellman-Ford algorithm, on the other hand, is a dynamic programming algorithm that repeatedly updates the estimated shortest path between nodes until it converges, and finally obtains the shortest path. These shortest path algorithms are often used in the calculation of intelligent routes in SDN to achieve efficient transmission and load balancing.

In addition, in SDN, the controller communicates with network devices through the OpenFlow protocol to dynamically configure the flow table, so as to achieve flexible management and optimization of network traffic. Compared with the traditional routing decision based on BGP (Border Gateway protocol), the centralized control of SDN can adjust the routing policy more flexibly and realize the fine control of network traffic. Telemetry data plays a key role in SDN and is closely related to the realization of intelligent network monitoring. By collecting real-time data from network devices, such as traffic, latency, packet loss rate and other metrics, telemetry can provide a comprehensive insight into the state of the network. SDN controllers can leverage this data to enable intelligent network monitoring, which allows real-time monitoring of network performance, health, and traffic patterns. Specifically, telemetry data helps SDN controllers identify changes in network topology, link status, and traffic distribution in real-time. By analyzing this data, the SDN controller can intelligently adjust network policies, such as dynamic route optimization, traffic adjustment, and fault recovery, to meet different QoS requirements and improve network performance and reliability.

6.3 Opportunities and Challenges Faced by SDN Under the Integration of Machine Learning

With the integration of machine learning, SDN will face new opportunities and new challenges. As mentioned in the previous chapter, intelligent technologies led by machine learning have made considerable contributions to intelligent network management and optimization of SDN networks and intelligent security protection. In contrast, the opportunities and challenges faced by SDN will be closely related to these two points. These opportunities include intelligent network management and optimization and intelligent security protection, as well as dynamic network optimization and intelligent application services. Seizing the opportunity will further enhance the performance of SDN and play a huge role in technology upgrading and employment in related fields. Relatively new technologies will also bring new challenges, such as the difficulty of technology convergence, data privacy security, and network model performance efficiency. The integration of machine learning and SDN needs to overcome the compatibility and interoperability between different technical systems and architectures, so the technology integration is not small. Data privacy security and network scale issues are two of the most common and common challenges in network systems: Machine learning is based on user data analysis and upgrade, where how to protect user data privacy is crucial. With the upgrading and expansion of SDN, network traffic continues to increase, and it will be increasingly difficult to ensure the performance and efficiency of machine learning.

In the future, intelligent technology in software-defined networks will develop in the following three aspects: First, intelligent network configuration and scheduling. Second, intelligent network operation and fault handling. Third, intelligent security protection mechanism. With the addition of machine learning and artificial intelligence technology, through the analysis and learning of massive data, the SDN network will tend to be adaptive and intelligent, and the network behavior and service quality will be adjusted according to the needs in different scenarios, so as to provide users with personalized intelligent application services.

An example of SDN intelligent network configuration is machine learning-based Traffic Engineering. In SDN networks, traffic engineering is an important network optimization technology. Traffic paths can be dynamically adjusted according to network traffic conditions and user requirements to improve network performance and resource utilization. The SDN controller collects data to train machine learning models, such as NNs and decision trees, to predict network traffic trends and identify network bottlenecks. According to the prediction results of the machine learning model, the SDN controller can intelligently adjust the path of traffic in the network to avoid congested nodes, reduce latency, improve bandwidth utilization, and so on. In addition, machine learning models can also analyze network traffic and device behavior, identify abnormal patterns and potential security threats, and achieve intelligent security protection purposes. For example, you can detect DDoS attacks, malware spread, unauthorized access, and more.

With the rise of 5G technology, the IoT, edge computing and other emerging fields in the same era, in order to successfully grasp the opportunity, intelligent technology and SDN network should be integrated with the development of emerging fields. The following will describe the relationship and integration of these three areas with SDN networking and machine learning.

Network slicing is an important feature in 5G networks that allows network operators to create multiple logically independent virtual network instances on the same physical network infrastructure based on different service requirements and service types. The network slicing technology in 5G network can divide network resources into multiple independent virtual networks, so that different services can enjoy customized network services. The combination of intelligent technology and SDN can realize intelligent management and optimization of network slices, dynamically adjust resource allocation according to service requirements, and improve network performance and user experience. Under the edge computing architecture of 5G network, the combination of intelligent technology and SDN can also realize intelligent edge computing resource scheduling and management, so as to support edge computing. The famous RL algorithm is the best example here.

Similarly, IoT involves a large number of device connections, and intelligent technology and SDN can realize the intelligent management and security protection of IoT devices. Monitor device behavior in real-time with machine learning models to identify anomalies and take automated action. For example, SVM, Decision Tree (Decision Tree) and Neural Network algorithms are used to build intelligent management and security protection systems to monitor the behavior of IoT devices in real-time and take automated actions. Ensure network security and stability.

It is worth noting that SDN algorithm plays a role in optimization and auxiliary control. In SDN network, centralized controllers can be used to manage and adjust policies in a unified manner. The machine learning algorithm can optimize the network strategy according to the different behavior patterns of devices in the SDN network, and improve the efficiency and security of device connection. SDN supports flexible traffic control and isolation based on flow tables, and can automatically isolate and restrict access to infected devices based on the detection results of machine learning algorithms to prevent security threats from spreading to the entire network.

In addition, in the edge computing environment, intelligent technology and SDN can realize resource scheduling and load balancing for edge nodes, reasonably allocate computing resources according to real-time data traffic and computing load, and improve computing efficiency. However, in the absence of SDN technology, it is difficult to dynamically adjust and optimize the utilization of edge node resources, which may lead to resource overload of some nodes and idle resources of other nodes, resulting in low resource utilization. The lack of SDN technology will also negatively affect the intelligent management and security protection control of network traffic, making the edge computing environment vulnerable to network attacks and security threats, affecting the stability and security of the system.

Next, we will further analyze and speculate the possible development trend of SDN from the aspect of future planning. As an important development direction of network technology, the analysis of its future planning can guide the direction of related technology research and development, help industry researchers better understand the development trend and focus of SDN technology, promote industrial application, help industry understand its development potential and application scenarios in different fields, and foresee possible challenges and risks. Develop coping strategies and measures in advance. Even in the best case, the development of SDN technology will be supported and guided by government policies, and the analysis of its future planning can provide decision-making reference for relevant government departments. This is why we will analyze the future planning of SDN.

SDN technology itself solves the two problems of network programmability and centralized control, and future planning will be based on them. We speculate that SDN will have more powerful network programming capabilities and cross-domain network programs and service offerings in the future.

SDN, as a network architecture, has shown a trend of combining with AI, and in the future it will be further combined with deep learning, RL, genetic algorithms, evolutionary algorithms and graph theory, etc., to support more diverse and complex network functions and services, including network virtual, network security and so on. At the same time, it will support open and standardized programming interfaces, making it easier for third-party developers to develop and integrate their own web applications.

As for cross-domain network programming and service provision, SDN has realized cross-domain network programming to a certain extent, but there are still some challenges. The SDN architecture allows multiple controllers to be deployed in different domains, and can also create, configure and manage cross-domain service links by defining service links. However, cross-domain network programming needs to solve cross-domain security policy coordination and performance optimization and other problems, but also unified development of programming interfaces and standards, so although the implementation of cross-domain network programming has made some progress, it still needs further research, which will become a direction of development and technology advancement in the future.

In traditional networks, traffic routing is usually handled by routing protocols on routers and switches, which can be distance vector protocols (such as RIP) or link state protocols (such as OSPF). These protocols select routing paths based on network topology and predefined policies.

RIP uses distance-based routing, where distance is a metric, usually the number of hops. Each router maintains a routing table that contains information about the distance to the destination network. Routers update these routing tables by periodically exchanging routing update messages.

To sum up, software-defined network will have both flexibility and manageability, which will make it have a broad application field in the future. For example, the flexibility of software-defined network will enable it to adapt to the dynamic change of cloud computing environment. In cloud computing and data center network, it can realize dynamic network resource allocation and optimization. At the same time, its manageability and flexibility will also enable it to support the needs of 5G networks.

In 5G networks, software-defined networks can realize the management and dynamic scheduling of network slices, and in edge computing environments, software-defined networks can realize the intelligent management and resource allocation of edge networks to reduce latency. SDN will also be widely used in the IoT and network security, providing efficient and flexible network support for different fields.

Chapter 7
Security in Wireless Communication

7.1 Secure Communications with Supervised Learning

Secure communication refers to the protection of communication content and the privacy and security of communication participants through various security mechanisms and technologies. In the field of communications, secure communication is crucial, especially in the age of the Internet. Traditional secure communication technologies typically rely on encryption algorithms and key exchange protocols to protect communication content from unauthorized access. However, with the increasing computational power of computers and advancements in cryptographic attack techniques, traditional encryption technologies may be vulnerable, thus necessitating more advanced security mechanisms to address these challenges.

Among the myriad of approaches being explored, the integration of supervised learning into security protocols represents a promising frontier in the quest for secure communications. Supervised learning, a cornerstone of machine learning, involves training an algorithm on a labeled dataset to enable it to make predictions or decisions without being explicitly programmed to perform the task. In the realm of wireless communication security, Supervised learning algorithms learn from data that encapsulate the characteristics of both secure and compromised communication channels. By analyzing patterns and anomalies in this data, the algorithms can distinguish between legitimate and malicious activities, thereby enhancing the security of wireless networks.

The practical applications of supervised learning in the field of Wireless Security are diverse. First, we will commence by exploring IDS. These systems capitalize on the capabilities of supervised learning algorithms, which undergo training on datasets inclusive of both ordinary network traffic and a variety of cyber-attack instances. Upon implementation, IDS can swiftly identify potential threats in real-time, providing network administrators with the tools to proactively address unauthorized access and counteract malicious activities. This proactive approach enhances the overall security posture of wireless networks.

H. Zhang and N. Yang, *Deep Learning in Wireless Communications*,
https://doi.org/10.1007/978-981-97-6314-6_7

Moreover, supervised learning is instrumental in Anomaly Detection within wireless networks. By grasping the typical operational parameters of a network, supervised learning models excel at pinpointing deviations from these norms, signaling a potential security breach or intrusion attempt.

Additionally, supervised learning plays a pivotal role in phishing attack prevention. With the rise of phishing attacks in wireless communication, supervised learning algorithms can be trained to detect telltale signs of phishing attempts, such as suspicious URLs and email content, effectively thwarting these attacks before they reach their targets.

Furthermore, supervised learning contributes to bolstering secure authentication protocols in wireless networks. By scrutinizing login patterns and user behavior, supervised learning models can discern between legitimate users and potential intruders, promptly blocking unauthorized access attempts and safeguarding both user data and network integrity.

Key techniques and algorithms of supervised learning are as follows: Several supervised learning algorithms are pivotal in enhancing wireless security, including:

- Decision trees: Used to model decisions and their possible consequences, making them suitable for classifying types of network attacks.
- Support Vector Machines (SVMs): Effective for high-dimensional data, SVMs are adept at distinguishing between benign and malicious traffic.
- Neural networks: With their ability to learn complex patterns, NNs are increasingly employed in detecting sophisticated cyber threats.

And supervised learning can be applied to user authentication through the following stages:

- Stage 1: Data Collection and preprocessing
  (1) Data collection
  Gathering Data: Collect data representing both normal and malicious network traffic. This data could include packet logs, network flow statistics, and user behavior metrics.
  Labeling: Each data point must be labeled as "normal" or "malicious" based on known outcomes. This step is crucial for supervised learning, as it provides the ground truth for training the model.
  (2) Preprocessing
  Feature extraction: Extract relevant features from the data that could indicate malicious activity. These features might include source / destination IP addresses, packet sizes, timestamps, protocol types, and payload contents.
  Normalization: Scale the features to a similar range to ensure no single feature dominates the model's learning process due to its scale.
  Data splitting: Divide the dataset into training, validation, and test sets. A common split ratio is seventy percent training, fifteen percent validation, and fifteen percent test.

- Stage 2: Model selection and training
  (1) Model selection
  Choose suitable supervised learning algorithms based on the nature of the data and the specific security task. Common choices include decision trees, SVM, and NNs.
  (2) Training
  Model training: Train the selected model on the training dataset by adjusting the model's parameters to minimize the difference between the predicted and actual labels.
  Validation: Use the validation set to tune hyperparameters and prevent overfitting. Techniques like cross-validation can be particularly useful here.
- Stage 3: Model evaluation testing: Evaluate the model's performance on the unseen test dataset to estimate how well it will generalize to new data.
  Performance metrics: Use metrics such as ACC, precision, recall, F1 score, and ROC-AUC to assess the model's effectiveness in detecting malicious activities.
- Stage 4: Deployment and real-time prediction deployment: Integrate the trained model into the network's security infrastructure, where it can analyze incoming traffic in real-time.
  Real-time prediction: As network traffic flows in, the model classifies it as normal or malicious, flagging potential threats for further investigation or automatic mitigation.
- Stage 5: Feedback loop and model updating Continuous Learning: The cyber threat landscape is constantly evolving. Regularly update the model with new data to ensure it remains effective against emerging threats.
  Feedback loop: Implement a feedback mechanism where the model's predictions are reviewed, and any misclassifications are corrected. Use this feedback to retrain the model, further enhancing its ACC and reliability.

Additionally, there exist advanced techniques and considerations that can enhance the effectiveness of this algorithmic flow (Fig. 7.1):

- Ensemble methods: These methods involve combining predictions from multiple models to enhance overall ACC and robustness, offering a more comprehensive approach to user authentication.
- Adversarial training: By integrating examples of adversarial attacks during the training process, the model can better recognize and defend against such threats, thereby increasing its resilience and security.
- Privacy-preserving techniques: Implementing techniques like federated learning allows models to be trained on decentralized data sources while preserving user privacy. This approach ensures that sensitive user information remains protected throughout the authentication process.

Challenges in implementing supervised learning for wireless security While the potential of supervised learning in enhancing wireless security is immense, several challenges must be addressed:

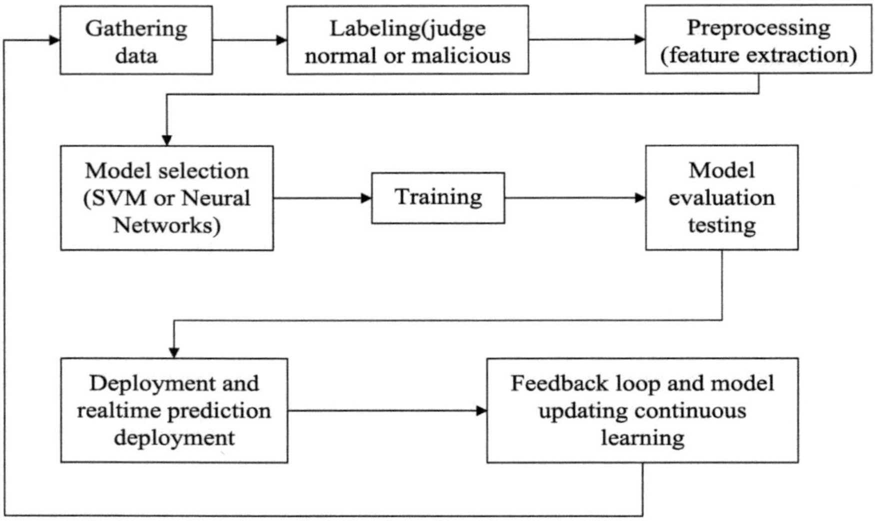

Fig. 7.1 The process of supervised learning be applied to user authentication

- Data quality and availability: The effectiveness of SL models is heavily dependent on the quality and diversity of the training data. Collecting comprehensive datasets that accurately represent the spectrum of potential cyber threats is challenging.
- Model complexity and overfitting: There is a delicate balance between model complexity and its generalizability. Overly complex models may overfit the training data, leading to poor performance on unseen data.
- Adversarial attacks: Cyber attackers continually evolve their strategies, potentially misleading SL models through crafted inputs designed to exploit weaknesses in the learning algorithm.

Future research in these areas is likely to focus on developing more resilient algorithms that can learn from smaller datasets, as well as enhancing the efficiency of these models to make them more suitable for real-time applications.

The future of secure communications with supervised learning is promising yet demands continuous innovation. Key areas for future research include:

- Federated learning: This approach enables models to learn from decentralized data sources without compromising user privacy, enhancing both security and privacy.
- Transfer learning: Leveraging pre-trained models on new, related tasks can reduce the need for extensive training data, addressing the challenge of data scarcity.
- Quantum machine learning: Quantum computing offers the potential to process complex algorithms more efficiently, paving the way for more advanced supervised learning models in wireless security.

And as wireless communication continues to permeate every aspect of modern life, the role of supervised learning in ensuring the security of these networks will only grow in importance.

In conclusion, the integration of supervised learning into wireless communication security represents a significant leap forward in the fight against cyber threats. By leveraging the predictive power of supervised learning, it is possible to develop more adaptive, efficient, and robust security measures. Despite the challenges, the ongoing advancements in machine learning technologies and methodologies hold great promise for creating a safer digital environment. As we continue to explore the vast potential of supervised learning, it is clear that its role in securing wireless communications will only become more pivotal in the years to come.

7.2 Federated Learning-Based Secure Communications

Federated learning, also known as federated machine learning, is a method proposed to address privacy concerns during joint model training. In this approach, each enterprise trains its own model. Upon completion of training, each enterprise uploads its model parameters to a central server, which can also be peer-to-peer. The central server combines these parameters, either by uploading gradients or updating its own parameters, and formulates new parameters using techniques like weighted averaging. These new parameters are then sent back to each enterprise, which deploys them to continue training. This iterative process continues until the model converges or other predefined conditions are met.

Unlike traditional methods that centralize training data, leading to potential single points of failure, sensitive data leakage, and significant overhead in collecting and storing training data, federated learning allows distributed mobile devices to co-train a global model using raw data while keeping this data on the mobile device.

In the communications technology industry, the Third Generation Partnership Programme (3GPP), a prominent standards development organization, has recognized and explored the use of machine learning and vast amounts of data as viable solutions to its challenges. The industry must investigate learning solutions capable of efficiently handling distributed datasets. Federated machine learning, an emerging decentralized approach, leverages on-device processing power and private data to train models in a decentralized manner, ensuring data stays where it is generated. While research in federated learning is still in its early stages, several challenges remain, especially in communication security.

Federated learning has a significant impact on communications security due to its distributed nature and the privacy-preserving techniques it employs. Here's how it relates to communications security:

1. Reduced data transmission: Federated learning minimizes the need to transmit raw data across networks. Instead of sending large volumes of sensitive data from individual devices to a central server, only model updates (which are typically much smaller) are communicated. This reduction in data transmission mitigates the risk

of interception or unauthorized access during communication, enhancing overall security.

2. Privacy preservation: By keeping data localized and performing computations on the device or at the edge, federated learning maintains data privacy. The emphasis on local training and transmitting model updates rather than raw data helps prevent sensitive information exposure during communication. This privacy-centric approach aligns with communication security principles by reducing the chances of data breaches or leaks during transmission.

3. Secure aggregation: Federated learning involves aggregating model updates from multiple devices at a central server or aggregator. Techniques used for secure aggregation, such as encryption or cryptographic protocols, ensure that these aggregated updates remain protected during transmission. This secure aggregation contributes to the overall communications security by safeguarding the integrity and confidentiality of the model updates.

4. Encrypted computation: Some federated learning approaches incorporate cryptographic techniques like homomorphic encryption or secure multiparty computation. These methods allow computations to be performed on encrypted data, ensuring that even during communication or collaborative learning, the data remains encrypted, thereby enhancing security.

Overall, federated learning intersects with communications security by emphasizing data privacy, minimizing data transmission, employing secure aggregation methods, and leveraging encryption techniques to safeguard information during communication and collaborative learning processes.

Federalism learning, or federated learning, is a machine learning approach that allows for training models across decentralized devices or servers while keeping data localized and private. In the context of wireless security, federated learning can address several issues:

1. Privacy concerns: Wireless networks often handle sensitive data. Federated learning enables model training without transferring raw data to a central server. Instead, models are trained locally on devices, preserving user privacy.

2. Data security: Transmitting data wirelessly can be vulnerable to interception or hacking. Federated learning reduces the risk by keeping data decentralized and localized, limiting exposure to potential breaches during transmission.

3. Resource efficiency: Wireless devices often have limited resources like battery power or bandwidth. Federated learning allows these devices to participate in model training without continuously transmitting large volumes of data, conserving resources.

4. Adaptability and localization: Wireless networks operate in diverse environments. Federated learning allows models to adapt to local conditions without relying on a centralized approach, leading to more contextually relevant security measures.

5. Collaborative threat detection: Federated learning enables multiple devices or network nodes to collaboratively train models to detect security threats. This distributed approach allows for a more comprehensive understanding of threats across the network without compromising individual data.

6. Continuous learning and adaptation: Wireless security threats evolve rapidly. Federated learning allows for continuous model updates across distributed devices, ensuring that security measures can adapt in real-time to emerging threats.

7. Regulatory compliance: Federated learning helps address regulatory concerns regarding data privacy and localization regulations. By keeping data local and minimizing data transfer, it can assist in compliance with various regional data protection laws.

However, implementing federated learning in wireless security also comes with its own set of challenges, such as synchronization of models across distributed devices, ensuring model integrity, and managing communication overhead. Additionally, maintaining the security of the federated learning framework itself is crucial to prevent attacks on the learning process.

Overall, while federated learning offers promising solutions for addressing several wireless security issues, its implementation requires careful consideration of technical, privacy, and security aspects to ensure its effectiveness.

Next, we will explain the role of federated learning in wireless network security from the first three aspects.

To address privacy security in wireless network communication, federated transfer learning (FTL) technique can be used. FTL is suitable for training machine learning models when the data set has different id Spaces and feature Spaces, and here's how it works:

1. Decentralized learning: Instead of pooling data into a central location, federated learning allows models to be trained across multiple decentralized devices or servers. Each device holds its own data locally.

2. Model distribution: A base model is initially created and distributed to individual devices. These devices then train the model further using their local data while keeping the model parameters on their premises.

3. Parameter updates: After training on local data, only the model updates (rather than raw data) are sent back to a central server or aggregator. This update contains the modifications made to the initial model during the local training process.

4. Aggregation: The central server aggregates the updates received from various devices to improve the global model. This aggregated update is then sent back to the devices, and the process iterates.

In a regression problem, the mean square error loss function is used to measure the distance from a sample point to a regression curve by minimizing the squared loss so that the sample point can better fit the regression curve. The smaller the value of the MSE function, the better ACC the predictive model has in describing the sample data. Due to the advantages of being parameter-free, computationally inexpensive, and having a clear physical meaning, MSE has become an excellent distance metric. Despite its weak performance in image and speech processing, MSE is still a criterion for evaluating signal quality, and it is often used as a model's empirical loss or an algorithm's performance metric in regression problems.

This process ensures that sensitive raw data never leaves the local devices, enhancing privacy. The central server typically only receives model updates or gradients, which are aggregated without exposing individual data points.

However, while federated learning helps with privacy, it's not foolproof. There are still potential vulnerabilities at different stages, such as inference attacks, model inversion attacks, or when adversaries gain access to multiple models' updates to reconstruct sensitive information. Techniques like differential privacy and encryption methods are often used in conjunction with federated learning to further enhance privacy protections.

Federated learning to deal with data security has the following key points:

1. Decentralization: Federalist or federated systems distribute data across various nodes or entities rather than centralizing it. This distribution minimizes the risk of a single point of failure and unauthorized access.

2. Access controls: These systems often employ robust access controls, ensuring that only authorized individuals or entities can access specific data or resources. This might involve authentication measures like multi-factor authentication, role-based access, and encryption.

3. Encryption: Data in transit and at rest is typically encrypted in federalist systems. This ensures that even if data is intercepted or accessed without authorization, it remains unreadable without the decryption keys.

4. Interoperability and standards: Federalist systems often adhere to standardized protocols and interoperability frameworks. This ensures that different components or entities within the system can communicate securely without compromising data integrity or security.

5. Monitoring and auditing: Continuous monitoring and auditing of the system help identify potential vulnerabilities or breaches. Logging and tracking activities within the system provide insights into any unauthorized access attempts or anomalies.

6. Regulatory compliance: Compliance with relevant data protection laws and regulations is a crucial aspect of federalist systems. This includes adherence to standards like GDPR, HIPAA, or other industry-specific regulations.

Homomorphic encryption technology plays an important role in the process of data security and the homomorphic encryption algorithm is divided into addition homomorphic algorithm, multiplication homomorphic algorithm and total homomorphic algorithm.

The additive homomorphic encryption algorithm is a subtype of homomorphic encryption that allows addition operations to be performed on encrypted data without the need to decrypt it first. This means that addition operations can be executed on encrypted data, and upon decryption, the result obtained matches the result of performing the addition operation on plaintext.

In additive homomorphic encryption, two encrypted messages (or an encrypted message and plaintext) can be added together to create a new encrypted message whose decryption result matches the sum of the corresponding plaintexts. This property is highly valuable in various applications, especially those requiring computation while maintaining data encryption, such as privacy-preserving cloud computing or secure data aggregation.

It's important to note that additive homomorphic encryption only supports addition operations and does not directly support other, more complex operations (such as multiplication). However, by employing various techniques and combinations, multiple addition operations can be combined to achieve a broader range of computations and functionalities.

Remember, while a federalist or federated system offers advantages in terms of scalability and distributed control, its effectiveness in ensuring data security relies heavily on the implementation of robust security measures across all participating nodes or entities.

Federated learning enhances resource efficiency in multiple ways:

1. Reduced data movement: Instead of transferring raw data to a central server, federated learning only shares model updates or aggregated information. This minimizes the amount of data transmitted, conserving bandwidth and reducing latency.

2. Localized computations: Devices or nodes perform model training locally with their data, eliminating the need to send data to a central location. This preserves data privacy and reduces the load on centralized servers, utilizing local resources more efficiently.

3. Decentralized training: By distributing the learning process across devices, federated learning decentralizes computation. This spreads the workload across the network, enhancing scalability and preventing bottlenecks on specific servers.

4. Edge computing utilization: Federated learning often leverages edge devices like smartphones or IoT devices for local training. This utilizes the computational power available at the network edge, optimizing resource utilization and minimizing reliance on central servers.

5. Incremental updates: Rather than retraining models from scratch with all available data, federated learning allows models to be updated incrementally. This reduces computational requirements for updates and enables continuous learning without extensive resources.

6. Scalability: Federated learning scales effectively with the addition of more devices. It accommodates a larger number of contributors without overloading central servers, making it suitable for large-scale applications.

Overall, federated learning's distributed approach to machine learning optimizes resource utilization, reduces data movement, enhances scalability, and safeguards privacy, making it an efficient framework for collaborative model training.

Next, we will discuss how federated learning can improve bandwidth efficiency. Federated Learning enhances bandwidth efficiency primarily through its decentralized learning approach, where model training happens locally on devices or at the edge rather than centralizing data on a server. Here's how it improves bandwidth efficiency:

Reduced Data Transmission:

1. Local model training: Instead of sending raw data to a central server, federated learning sends only model updates or gradients after training on local data. These updates are much smaller in size compared to the entire dataset.

2. Lower communication overhead: Communication occurs between the local devices/edge nodes and the central server selectively, reducing the need for continuous data transmission.

Bandwidth Optimization:

1. Dynamic model updates: Federated Learning allows for more flexible scheduling of updates, which can be based on connectivity status, available bandwidth, or device activity. This optimizes bandwidth usage.

2. Local computation: By leveraging local computation, only model updates are transmitted, reducing the need for constant high-bandwidth connections.

Edge Computing Advantages:

1. Edge servers as learning centers: Utilizing edge devices or servers for learning reduces the need for continuous data transfers to a centralized cloud, conserving bandwidth.

2. Real-Time learning: Federated Learning enables devices to learn and adapt in real-time at the edge, reducing the necessity for frequent model synchronization, which in turn saves bandwidth.

A variety of algorithms and techniques can help with bandwidth optimization, and arithmetic coding is one of them.

Arithmetic coding is a commonly used lossless data compression technique that uses a probability distribution to encode data. The core idea of arithmetic coding is to map the entire data sequence to a sub-interval on an interval [0, 1), thus representing the entire data sequence as a floating point number. The length of the encoding is variable and depends on the probability distribution of each symbol (character or data element) over the entire sequence.

This process keeps repeating, updating the interval and shrinking it for each symbol of the input. Ultimately, the encoding is any value of the interval. The decoding process then reduces the original data based on the relationship between intervals and probabilities.

Note that in order to accurately represent floating point numbers, it may be necessary to use high-precision arithmetic operations. Arithmetic coding can theoretically achieve very high compression rates, but in practice it may be affected by computational ACC, decoding complexity, and other factors.

Federated learning optimizes bandwidth by minimizing the amount of data transmitted, focusing on model updates rather than raw data, preserving privacy by keeping data local, and leveraging edge computing for distributed learning. This approach reduces network congestion and bandwidth consumption, making it more efficient for training machine learning models across distributed devices or servers.

The connection between federated learning and communication security lies in federated learning's capacity to address critical security challenges in wireless networks. Its decentralized, privacy-preserving, and adaptive nature not only safeguards sensitive data but also fortifies communication systems against adversarial threats, ensuring efficient resource utilization and facilitating continual learning without compromising security or privacy. This symbiotic relationship presents a promising avenue for the evolution of secure wireless communication infrastructures.

In the realm of wireless communication, the symbiosis between federated learning and communication security heralds a transformative era. Federated learning's decentralized, privacy-centric approach not only safeguards sensitive data but also fortifies communication systems against adversarial threats. Its ability to optimize resource utilization, ensure continual learning, and bolster network resilience marks a paradigm shift towards secure wireless communication infrastructures.

As wireless networks navigate evolving threats and stringent privacy concerns, federated learning emerges as a beacon of innovation. Its capacity to harmonize robust security measures with efficient data processing without compromising privacy presents a promising avenue for the future. By harnessing the power of distributed learning, wireless networks can evolve into resilient, adaptable, and secure ecosystems, ensuring the integrity of data transmission while enabling continual improvement in an increasingly interconnected world.

In this landscape of evolving communication technologies, federated learning stands not just as a solution but as a catalyst for a decentralized, secure future where communication security and data privacy coexist harmoniously, shaping the next generation of wireless networks.

Chapter 8
6G Driving Applications with Deep Learning

8.1 Application Scenarios and Challenges

In the rapid development of information technology, the development cycle of the fifth generation of new radio program has reached preliminary maturity, has a higher transmission speed, lower delay and larger capacity, but there is still a lot of room for progress. 5G's inability to deliver a fully automated and intelligent network that delivers everything as a service and provides a fully immersive experience will not meet the demand for emerging and automated systems over the next decade. And it largely ignores the integration of communication, intelligence, sensing, control, and computing functions, failing to meet people's expectations of supporting IoT applications. 5G will reach its limits in the next decade, which is a huge departure from people's expectations that the sixth generation of mobile communications will be born.

Compared with 5G technology, 6G combines all the features of the past, such as network densification, high throughput, high reliability, low energy consumption and large-scale connectivity, which brings the hardware foundation for the implementation of the IoT and makes more applications possible. At the same time, there are greater breakthroughs in transmission speed, delay, capacity and other aspects, and it will also have higher density connection capabilities and more secure security mechanisms, which will bring users a more comfortable experience. With the gradual maturity of 6G technology and the gradual development of commercial applications, we will usher in a new chapter of the digital era, and an unprecedented intelligent world is coming.

This chapter aims to explore the future of 6G applications and stand at the forefront of innovation, focusing on the application of 6G in intelligent transportation, robotics, medical fields, and other aspects, as well as the challenges that 6G applications will face. By understanding the application of 6G in different fields, explore its potential value in improving people's quality of life and promoting the development of various industries.

H. Zhang and N. Yang, *Deep Learning in Wireless Communications*,
https://doi.org/10.1007/978-981-97-6314-6_8

1. Intelligent transportation

6G technology can bring higher data processing speed and capacity to intelligent transportation systems, so as to better cope with the processing needs of large amounts of data. By connecting traffic equipment, sensors and vehicles to the 6G network, and combining with AI, it can collect and transmit a large amount of traffic data in real-time, such as vehicle location, speed, driving route and other information, AI technology can carry out real-time identification of vehicles and personnel, behavior analysis and anomaly detection, and quickly discover and alarm traffic accidents, violations and other situations. This will help the intelligent transportation system to more accurately monitor the driving state of the vehicle, predict and solve the traffic result congestion, accidents and other problems in advance. In addition, 6G technology will also improve the delay problem of intelligent transportation systems and provide more efficient communication connections. In the 6G network environment, the communication between intelligent transportation devices will achieve almost no delay transmission, which will provide more timely and accurate instructions and information for intelligent vehicles and traffic equipment. Smart vehicles can use 6G technology to achieve faster decision-making and reaction, thereby reducing the incidence of traffic accidents. At the same time, the intelligent transportation system can also use the 6G network to achieve collaborative communication between vehicles, optimize the control of intelligent traffic lights, and improve traffic efficiency. Combined with AI, it can also realize intelligent signal control and intersection optimization, automatically adjust the timing of signal lights according to real-time traffic conditions, reduce traffic congestion and driving fuel consumption, and improve traffic efficiency and environmental friendliness.

2. IoT

6G technology will enable high-speed connection and super-capacity data transmission of IoT devices, providing a solid foundation for the development of IoT applications. The combination of 6G networks and AI will be able to support the simultaneous connection of more devices and provide high-speed and stable data transmission channels. This will enable IoT devices to collect, transmit and process a large amount of data more quickly, and by deploying AI algorithms and data processing capabilities on edge devices, IoT devices can quickly collect and analyze a large amount of real-time data, and make intelligent decisions based on this data, bringing a richer intelligent experience to our lives. At the same time, it can also make the IoT system have lower latency and higher reliability, and achieve more accurate interconnection. The latency of 6G networks is almost negligible, making communication between IoT devices timely and real-time. This will enhance the collaboration and interaction between IoT devices and improve the reliability and stability of the system. With the advancement and popularization of 6G technology, the IoT will provide more possibilities for our lives, allowing us to achieve deeper connectivity and interaction with smart devices and objects.

3. Robots

6G technology will provide more high-speed data transmission and processing capabilities for robot applications, and achieve more advanced human-computer interaction, decision-making and autonomous action capabilities through AI algorithm analysis and decision-making. Robots need to process a large amount of data in the process of perception, decision making and execution, and 6G technology has extremely high data transmission speed and processing capability, which can realize real-time data interaction and intelligent decision-making. This will enable robots to acquire and process large amounts of information more quickly and accurately, improving their intelligence and autonomy. At the same time, the combination of 6G technology and AI will provide lower latency and more stable connections for robot applications, achieving more accurate remote control and collaboration. The 6G network builds a communication environment with extremely low delay, and the robot can realize almost real-time interaction with the human operator through the 6G network, further improving the ACC and flexibility of the robot's remote control. In addition, 6G technology will also support collaborative work and communication between robots, so that robots can better collaborate with each other to complete complex tasks. 6G will achieve a breakthrough in data processing and transmission of robots, making them more intelligent and autonomous. The stable connection and remote collaboration technology will promote the remote control and collaboration ability of robots, and bring a broader development space for our production, service and life.

4. Virtual and AR

6G technology will provide higher-speed data transmission and lower latency for VR and AR, allowing users to experience virtual and AR content more smoothly and realistically. With the support of the 6G network, users can quickly download high-resolution VR and AR content and enjoy an immersive experience without waiting. At the same time, ultra-low latency will make user interaction in virtual and AR environments more natural and real-time, further enhancing the user experience. The combination of AI can also generate personalized AR and VR content in real-time based on an individual's preferences, needs and environment through the analysis of user behavior and environment. This means that users can enjoy a more customized and personalized experience, increasing user engagement and satisfaction. The large capacity of 6G technology will provide more possibilities for content creation and interaction in VR and AR. In the 6G era, VR and AR can support richer and more complex virtual and AR content, making the application scenarios of virtual and AR more extensive. Through the support of 6G network, users can obtain and transmit virtual and AR content more quickly and accurately, expanding the application boundaries of VR and AR. For example, in the commercial field, through 6G technology, users can experience virtual shopping in real time, feel the shopping experience similar to that of physical stores, and improve the convenience and satisfaction of shopping.

5. Medical treatment

The ultra-high speed and ultra-low latency of 6G will make the transmission of medical data faster and more reliable. Healthcare organizations can transfer patients' medical data to the cloud in real-time through 6G networks, and doctors and specialists can access and analyze this data remotely to provide more accurate and timely diagnosis and treatment options. At the same time, ultra-low latency will also make remote surgery and remote consultation possible, and doctors can communicate and operate with patients in real-time through 6G technology, providing patients with accurate medical services. The application of 6G technology will greatly shorten the waiting time of patients and improve the utilization efficiency of medical resources. The large capacity of 6G will provide more support for big data analysis and artificial intelligence applications in the medical field. The collection and analysis of medical data is the key to improving the quality and efficiency of medical care. Through 6G technology, medical institutions can connect to massive medical databases and intelligent systems to achieve in-depth analysis and mining of medical data, while the application of AI technology can help doctors make disease prediction, diagnosis and treatment plans. This will provide more data support and decision-making basis for doctors and researchers, and promote innovation and progress in medical research and clinical practice. 6G technology will also open up greater space for the application of AI in the medical field, through machine learning and automation technology, to help doctors achieve more accurate and rapid diagnosis, and provide personalized treatment plans for patients.

6. The tactile Internet

The ultra-low latency and high-speed transmission capabilities of 6G technology enable tactile Internet to achieve almost real-time touch transmission. With 6G networks, we can touch, manipulate and sense remote objects or objects in real-time through haptic devices. This will make remote operations and VR experiences more real and immersive. In the field of VR, users can perceive the tactile feedback in the virtual environment in real-time through the 6G network to further enhance the immersion of the virtual experience. The large capacity of 6G technology will enable the tactile Internet to handle more touch data and higher precision touch information. Haptic Internet is not only a simple haptic transmission, but also the collection, analysis and feedback of haptic data. With the support of 6G network, the tactile device can collect more tactile data and realize fine perception of tactile details such as user gestures, pressure and temperature.

With the continuous progress and innovation of science and technology, the combination of AI and 6G communication has begun to show great potential. This combination has created unprecedented opportunities and challenges for various industries, including security and privacy issues, the difficulty of technology integration, and social acceptance.

1. Bandwidth and connectivity: a key goal of 6G technology is to provide higher transmission speeds and lower latency. However, transferring large amounts of data to AI systems, as well as returning results from AI systems, requires higher bandwidth and more stable connections. Ensuring the coverage and stability of 6G networks will be a challenge.

2. Privacy and security: AI systems rely on large amounts of data and generate a lot of personal information. Combined with 6G networks, the transmission and storage of this data will involve more nodes and cloud services. Ensuring the protection of privacy and security to prevent data breaches and malicious attacks is an important challenge.

3. Computing power and energy consumption: AI systems require powerful computing power to perform highly complex calculations. At the same time, in order to ensure the efficient energy consumption of 6G networks, a balance needs to be struck between AI algorithms and 6G technology. How to achieve high performance computing while keeping energy consumption low is a challenge.

4. Data management and standardization: In the combined environment of 6G and AI, a large amount of data will be collected, transmitted and analyzed. How to effectively manage and process this data, as well as establish uniform data standards and exchange formats, will be an important challenge.

5. Legal and ethical issues: The application of AI in 6G networks will raise a series of legal and ethical issues. For example, issues related to privacy, data security, and algorithmic bias. Developing regulations and ethical guidelines to balance innovation with protecting user rights is a challenge.

Although the combination of 6G and AI faces many challenges, it is believed that these challenges will be gradually overcome with the development of technology and the promotion of innovation. We look forward to the arrival of 6G, which will bring new opportunities and possibilities for the smart industry, smart cities and personal life. At the same time, we also look forward to the deeper application of AI technology in 6G networks to provide strong support for intelligent and efficient communication.

8.2 Enabling Deep Learning Technologies

Deep learning is a subfield of machine learning that builds and trains artificial neural network models by mimicking the way neurons in the human brain are connected. The goal of deep learning is to utilize large amounts of data and computational resources, automatically learn feature representations, and extract high-level abstract features from the data through hierarchical structures.

The background of deep learning can be traced back to artificial neural network research in the 1950s and 1960s, but the development of deep learning was limited by the limitations of computing power and the lack of large-scale data sets at the time. Only in recent years, with the rapid development of computing hardware and the explosive growth of Internet data, has deep learning technology been able to rise rapidly. In the past few years, deep learning has made major breakthroughs in many fields, including computer vision, natural language processing, speech recognition, and more. The advantage of deep learning is that it can automatically learn feature representations without manually extracting features, and it has strong generalization ability and can handle complex non-linear relationships. This makes deep learning an important technical foundation in the field of AI and has achieved remarkable results in many practical applications.

Deep learning operates on several core concepts and ideas:

1. Feedforward neural network: it is the most basic artificial neural network structure, composed of input layer, hidden layer and output layer, information from the input layer through a series of layers, and finally output the prediction result.

2. Backpropagation: it is the main algorithm for training NNs, adjusting the network parameters by calculating the output error to minimize the gap between the predicted results and the actual values.

3. Gradient descent: it is the basis of backpropagation algorithms and continuously optimizes the model by updating the parameters by calculating the gradient of the error function for each parameter.

Deep learning also involves some other important concepts and techniques, such as activation function, model regularization, batch normalization, etc. The use of these techniques and strategies can improve the effectiveness and robustness of the model.

The core idea of deep learning is to learn the feature representation of data layer by layer through a multi-level neural network structure to obtain higher-level abstract features. This hierarchical structure can be trained and optimized by backpropagation algorithm and gradient descent optimization algorithm to achieve accurate prediction and efficient processing of complex tasks.

Compared with traditional machine learning algorithms, deep learning has the following significant differences:

1. Feature extraction: Traditional machine learning algorithms usually need to manually extract and select features, while deep learning can automatically learn feature representations through hierarchical structures without manually extracting features, thus reducing the burden of feature engineering.

2. Model complexity: DL builds DNN models to represent complex nonlinear relationships with greater expressiveness. Traditional machine learning algorithms, such as SVMS and decision trees, are often used for shallow models that struggle to handle large-scale data and complex tasks.

3. Data requirements: DL has a high demand for large-scale data sets, and trains models through large amounts of data to improve the ACC and generalization ability of models. Traditional machine learning algorithms require relatively small amounts of data.

DL has a wide range of applications in various fields. Here are some examples of deep learning in different fields:

1. Computer vision: DL has been a huge success in computer vision. Among them, deep CNN performs well on tasks such as image classification, object detection and image generation. For example, through deep learning, applications such as face recognition, image semantic segmentation, and object recognition can be realized.

2. Natural language processing: DL also has important applications in the field of natural language processing. For example, using models such as RNN and LSTM, tasks such as language modeling, machine translation, text generation and sentiment analysis can be implemented.

3. Speech recognition: DL has made a breakthrough in the field of speech recognition. Through the combination of DNN and acoustic model, high ACC of speech

recognition can be achieved, and it is widely used in voice assistants, smart phone systems, etc.

4. Health care: DL is widely used in the field of health care, such as medical image analysis, disease diagnosis and prediction, genomics research, etc. DL can learn large amounts of medical data to extract and identify information such as medical images, image markers, and pathological analyses.

5. Financial sector: DL is widely used in the financial sector, such as stock forecasting, fraud detection and risk assessment. DL can provide accurate predictions and decision support by learning patterns and trends in financial data.

6. Autonomous driving: DL plays an important role in the field of autonomous driving. Through deep learning algorithms, functions such as vehicle perception, target detection and behavior prediction can be realized, providing critical decision-making capabilities for autonomous driving systems.

As a powerful machine learning method, DL has achieved remarkable results in several fields, but there are still some challenges and directions for future development. Here are some common challenges and the way forward:

1. Data requirements and labeling difficulties: DL usually requires a large amount of labeled data for training, and obtaining and labeling large-scale data sets is a time-consuming and laborious task. Therefore, how to effectively train when data is scarce or no labeled data remains a challenge.

2. Model interpretation and interpretability: DL models are often black-box models, and their complexity makes it difficult to explain how the model makes its predictions. In some application scenarios, such as healthcare and finance, the interpretability and explainability of the model are critical. Therefore, how to improve the interpretability of deep learning models is a future research direction.

3. Model generalization and avoiding overfitting: DL models are prone to overfitting on training sets, resulting in weak generalization ability on new data. How to design better regularization techniques and effective model selection methods to improve the generalization ability of models is an important problem in deep learning research.

4. Long-term memory and reasoning ability: Current deep learning models still have limitations in long-term memory and reasoning ability, especially for complex serial data and time series tasks. How to design deep learning models that can capture long-term dependencies and have stronger reasoning ability is an important direction for future development.

5. Hardware and computational resource constraints: The training and reasoning of deep learning models often require a large amount of computational resources, including computational power and storage capacity. How to design more efficient algorithms, develop more advanced hardware technologies, and utilize distributed computing methods to meet the demand for large-scale DL models is the future direction.

6. Incorporate other learning methods: although deep learning has had a lot of success, it is not suitable for every problem and task. Combining other machine learning methods, such as traditional machine learning algorithms and RL, to develop

a more comprehensive and integrated learning framework is the future development direction of deep learning.

Overall, while deep learning continues to grow, it still faces many challenges. By continuously addressing these challenges, improving deep learning models and algorithms, and incorporating other learning methods, deep learning is expected to achieve wider applications and more groundbreaking progress in the future.

8.3 New Paradigm Shifts

The construction of sea, land and air integrated communication network needs to rely on the support of intelligent technology, and then combined with wireless communication technology, in order to achieve efficient communication on a global scale. Intelligent technology makes communication network more intelligent and automated, and wireless communication technology provides flexible and convenient means of communication.

Although the combination of 6G and AI faces many challenges, it is believed that these challenges will be gradually overcome with the development of technology and the promotion of innovation. We look forward to the arrival of 6G, which will bring new opportunities and possibilities for the smart industry, smart cities and personal life. At the same time, we also look forward to the deeper application of AI technology in 6G networks to provide strong support for intelligent and efficient communication.

In the future, the sea, land and air integrated communication network will continue to improve the communication speed and reliability, meet the growing communication needs, and further realize the integration of different networks, including cellular networks, satellite networks, ground networks, etc., to achieve global coverage and seamless switching, providing integrated communication services across the sea, land and air. At the same time, it will also support the interconnection of different types of devices, so that it can be combined with the IoT technology to support the interconnection of different types of devices. This means that all kinds of devices, such as smartphones, smart home devices, intelligent vehicles, etc., can be connected and communicated through the integrated communication network of sea, land and air. For example, people can realize real-time video calls, telemedicine, smart home control and other applications on a global scale through the integrated communication network of sea, land and air. For enterprises and institutions, through the integrated communication network of sea, land and air, logistics companies can grasp the location, transportation status and traffic conditions of goods in real-time, so as to improve transportation efficiency and ACC. The integrated communication network of sea, land and air can also support rapid rescue and command in emergency situations, improving rescue efficiency and response speed.

As a branch of AI, machine learning will have an important impact on the process of 6G technology to realize the interconnection of integrated communication between sea, land and air. For example, machine learning can be applied to network optimization and resource scheduling in integrated sea, land and air communication networks. By analyzing large-scale communication data and network status information through machine learning algorithms, network resource configuration and scheduling policies can be optimized to improve network performance and user experience. Second, machine learning can be used for wireless resource management and intelligent connection management in 6G technology. 6G technology will introduce the use of more spectrum resources and more complex network topology to achieve integrated communication connections between sea, land and air, and machine learning can help realize intelligent management and optimization of wireless channels to improve spectrum utilization efficiency and network capacity. In addition to this, machine learning can also be used for intelligent edge computing and cybersecurity in 6G technology. 6G technology will support large-scale edge computing and the connection of IoT devices in sea, land and air integrated communication networks, and machine learning can be used for real-time data analysis and processing to achieve intelligent decision-making and optimization of edge computing. In logistics and emergency relief–as mentioned earlier: for logistics, machine learning can optimize the route planning of logistics transportation by analyzing historical transportation data, traffic conditions, weather conditions and other factors to predict the best route. Through machine learning algorithms, historical sales data, market trends and other relevant data can be analyzed to predict changes in logistics demand. For emergency rescue, machine learning can be used to analyze large-scale data, such as real-time sensor data, traffic information, etc., to help quickly respond to emergency rescue events, and quickly determine the best rescue plan and action plan based on the analysis.

Sub-6 GHz, millimeter wave, and terahertz bands complement each other in 6G technology to meet future demands for high-speed, high-capacity, low-latency wireless communications. The Sub-6 GHz band is the key band for the current mainstream mobile communication technologies (such as 2G, 3G, 4G and 5G), and it will continue to be used for 6G technology's wide coverage network, supporting wide-area transmission and long-distance communication, suitable for densely populated areas and vast geographical areas. The application of millimeter wave and terahertz band is still in the research and development stage, and the millimeter wave band will play an important role in the future 6G technology, providing high-speed, high-capacity data transmission. Despite the short transmission distance of millimeter wave, its large amount of idle spectrum can be used efficiently. The terahertz band may become a new communication band in 6G technology to support higher data transmission rates and a wider range of application scenarios, such as VR, AR, intelligent transportation and remote surgery.

This process is often affected by many factors such as climate and obstacles, which may lead to performance degradation or unstable signal transmission. However, machine learning can make predictions and optimizations by learning from this data, thereby improving channel quality and input stability, reducing data transmission errors and latency. On the one hand, machine learning can learn and predict signal transmission conditions under the influence of different climatic conditions and obstacles by analyzing a large amount of meteorological data, barrier distribution and

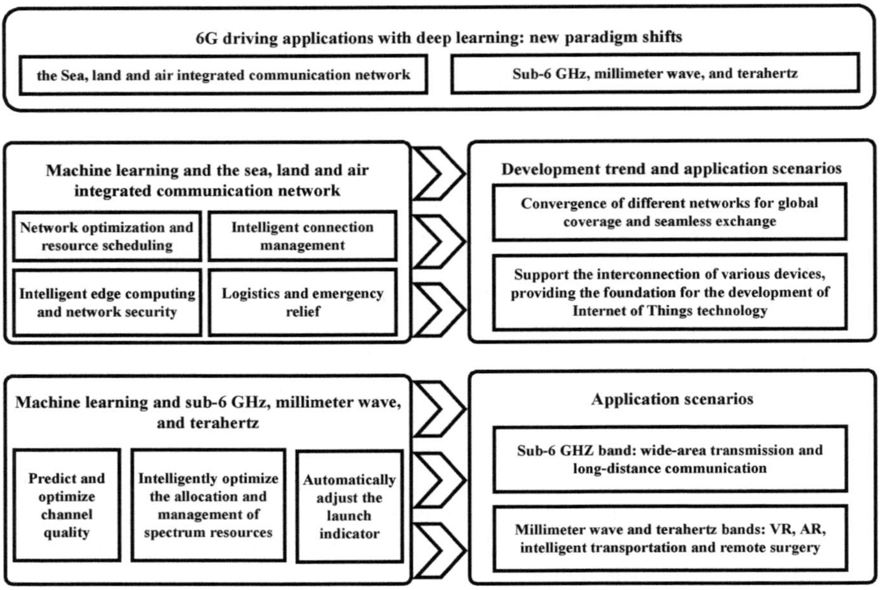

Fig. 8.1 Machine learning boosts new paradigm shifts in 6G

transmission performance data in actual scenarios. In this way, the communication system can automatically adopt corresponding adjustment strategies according to the prediction results of machine learning, such as dynamically adjusting the transmission power, changing the transmission Angle or choosing a more suitable communication frequency band, so as to improve the signal transmission quality and stability. On the other hand, since wireless spectrum resources are limited, machine learning can be applied to spectrum allocation and management to achieve more efficient spectrum utilization. By analyzing and learning from large amounts of data, machine learning can automatically optimize the allocation of spectrum resources, identify potential idle spectrum segments, and make adjustments to dynamic frequency and power allocation based on real-time demand. This can avoid unnecessary spectrum conflicts, improve the efficiency of spectrum utilization, and provide more bandwidth resources for the communication system to support larger data transmission and higher rate communication (Fig. 8.1).

MIX
Papier aus verantwortungsvollen Quellen
Paper from responsible sources
FSC® C105338

If you have any concerns about our products,
you can contact us on
ProductSafety@springernature.com

In case Publisher is established outside the EU,
the EU authorized representative is:
Springer Nature Customer Service Center GmbH
Europaplatz 3, 69115 Heidelberg, Germany

Printed by Libri Plureos GmbH
in Hamburg, Germany